THE HUMAN USE OF HUMAN BEINGS

CYBERNETICS AND SOCIETY

NORBERT WIENER

With a new Introduction by
Steve J. Heims

FA^B

'an association in which the free development of each is the
condition of the free development of all'

FREE ASSOCIATION BOOKS / LONDON / 1989

Published in Great Britain 1989 by
Free Association Books
26 Freegrove Road
London N7 9RQ

First published 1950; 1954, Houghton Mifflin
Copyright, 1950, 1954 by Norbert Wiener
Introduction © Steve J. Heims 1989

British Library Cataloguing in Publication Data

Wiener, Norbert, *1894–1964*
The human use of human beings: cybernetics
and society
1. Cybernetics. Sociological perspectives
I. Title
306'.46

ISBN 1-85343-075-7

Printed and bound in Great Britain by
Bookcraft, Midsomer Norton, Avon

KING ALFRED'S COLLEGE
WINCHESTER

To be returned on or before the day marked
below, subject to recall

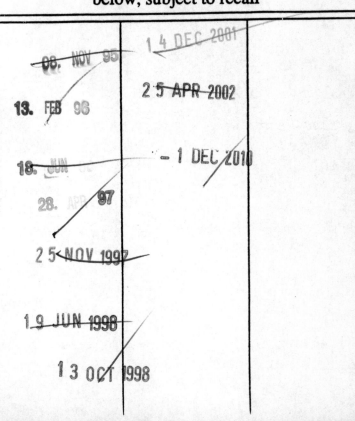

To the memory of my father
LEO WIENER
formerly Professor of Slavic Languages
at Harvard University
my closest mentor and dearest antagonist

ACKNOWLEDGEMENTS

Part of a chapter has already appeared in the *Philosophy of Science.* The author wishes to acknowledge permission which the publisher of this journal has given him to reprint the material.

CONTENTS

BIOGRAPHICAL NOTES

NORBERT WIENER, born in 1894, was educated at Tufts College, Massachusetts, and Harvard University, Massachusetts, where he received his Ph.D. at the age of nineteen. He continued his studies at Cornell, Columbia, in England at Cambridge University, then at Göttingen and Copenhagen. He taught at Harvard and the University of Maine and in 1919 joined the staff of the Massachusetts Institute of Technology, where he was Professor of Mathematics. He was joint recipient of the Bocher Prize of the American Mathematical Society in 1933, and in 1936 was one of the seven American delegates to the International Congress of Mathematicians in Oslo. Dr Wiener served as Research Professor of Mathematics at the National Tsing Hua University in Peking in 1935–36, while on leave from MIT. During World War II he developed improvements in radar and Navy projectiles and devised a method of solving problems of fire control.

In the years after World War II Wiener worked with the Mexican physiologist Arturo Rosenblueth on problems in biology, and formulated the set of ideas spanning several disciplines which came to be known as 'cybernetics'. He worked with engineers and medical doctors to develop devices that could replace a lost sensory mode. He analysed some non-linear mathematical problems and, with Armand Siegel, reformulated quantum theory as a stochastic process. He also became an articulate commentator on the social implications of science and technology. In 1964 Wiener was recipient of the US National Medal of Science.

His published works include *The Fourier Integral and Certain of Its Applications* (1933); *Cybernetics* (1948); *Extrapolation and Interpolation and Smoothing of Stationary Time Series with Engineering Applications* (1949); the first volume of an autobiography, *Ex-Prodigy: My Childhood and Youth* (1953); *Tempter* (1959); and *God and Golem* (1964). Wiener's published articles have been assembled and edited by P. Masani and republished in four volumes as *Norbert Wiener: Collected Works* (1985).

STEVE J. HEIMS received his doctorate in physics from Stanford University, California. He engaged in research in the branch of physics known as statistical mechanics and taught at several North American universities. In recent years he has devoted himself to studying various contexts of scientific work: social, philosophical, political and technological. He is the author of *John von Neumann and Norbert Wiener: From Mathematics to the Technologies of Life and Death* (MIT Press, 1980). Currently he is writing a book dealing with the characteristics of social studies in the USA during the decade following World War II.

INTRODUCTION

Steve J. Heims

G.H. Hardy, the Cambridge mathematician and author of *A Mathematician's Apology*, reflecting on the value of mathematics, insisted that it is a 'harmless and innocent occupation'. 'Real mathematics has no effects on war', he explained in a book for the general public in 1940. 'No one has yet discovered any warlike purpose to be served by the theory of numbers or relativity . . . A real mathematician has his conscience clear.' Yet, in fact, at that time physicists were already actively engaged in experiments converting matter into energy (a possibility implied by the Theory of Relativity) in anticipation of building an atomic bomb. Of the younger generation which he taught, Hardy wrote, 'I have helped to train other mathematicians, but mathematicians of the same kind as myself, and their work has been, so far at any rate as I have helped them to it, as useless as my own . . . '

Norbert Wiener took issue with his mentor. He thought Hardy's attitude to be 'pure escapism', noted that the ideas of number theory are applied in electrical engineering, and that 'no matter how innocent he may be in his inner soul and in his motivation, the effective mathematician is likely to be a powerful factor in changing the face of society. Thus he is really as dangerous as a potential armourer of the new scientific war of the future.' The neat separation of pure and applied mathematics is only a mathematician's self-serving illusion.

Wiener came to address the alternative to innocence – namely, taking responsibility. After he himself had during World War II worked on a mathematical theory of prediction intended to enhance the effectiveness of anti-aircraft fire, and developed a powerful statistical theory of communication which would put modern communication engineering on a rigorous mathematical footing, any pretence of harmlessness was out of the question for him. From the time of the end of the war until his death in 1964, Wiener applied his penetrating and innovative mind to identifying and elaborating on a relation of high technology to people which is benign or, in his words, to the human – rather than the inhuman – use of human beings. In doing so during the years when the cold war was raging in the United States, he found an audience among the generally educated public. However, most of his scientific colleagues – offended or embarrassed by Wiener's views and especially by his open refusal to engage in any more work related to the military – saw him as an eccentric at best and certainly not to be taken seriously except in his undeniably brilliant, strictly mathematical, researches. Albert Einstein, who regarded Wiener's attitude towards the military as exemplary, was in those days similarly made light of as unschooled in political matters.

Undaunted, Wiener proceeded to construct a practical and comprehensive attitude towards technology rooted in his basic philosophical outlook, and presented it in lucid language. For him technologies were viewed not so much as applied science, but rather as applied social and moral philosophy. Others have been critical of technological developments and seen the industrial revolution as a mixed blessing. Unlike most of these critics, Wiener was simultaneously an irrepressibly original non-stop thinker in mathematics, the sciences and high technology *and* equally an imaginative critic from a social, historical and ethical perspective of the uses of his own and his colleagues' handiwork. Because he gave rather unchecked rein to both of these inclinations, Wiener's writings generate a particular tension and have a special fascination.

Now, four decades later, we see that the tenor of his comments on science, technology and society were on the whole prophetic and ahead of his time. In the intervening years his subject matter, arising out of the tension between technical fascination and social conscience, has become a respectable topic for research and scholarship. Even leading universities have caught up with it and created courses of study and academic departments with names such as 'science studies', 'technology studies' or 'science, technology and society'. His prediction of an imminent 'communication revolution' in which 'the message' would be a pivotal notion, and the associated technological developments would be in the area of communication, computation and organization, was clear-sighted indeed.

The interrelation between science and society via technologies is only one of the two themes underlying *The Human Use of Human Beings*. The other derives as much from Wiener's personal philosophy as from theoretical physics. Although he was a mathematician, his personal philosophy was rooted in existentialism, rather than in the formal-logical analytical philosophy so prominent in his day and associated with the names of Russell, Moore, Ramsey, Wittgenstein and Ayer. For Wiener life entailed struggle, but it was not the class struggle as a means to social progress emphasized by Marxists, nor was it identical with the conflict Freud saw between the individual and society. In his own words:

We are swimming upstream against a great torrent of disorganiza-tion, which tends to reduce everything to the heat death of equilibrium and sameness described in the second law of thermo-dynamics. What Maxwell, Bolzmann and Gibbs meant by this heat death in physics has a counterpart in the ethic of Kierkegaard, who pointed out that we live in a chaotic moral universe. In this, our main obligation is to establish arbitrary enclaves of order and system. These enclaves will not remain there indefinitely by any momentum of their own after we have once established them . . . We are not fighting for a definitive victory in the indefinite future. It is the greatest possible victory to be, to continue to be, and to have been . . . This is no defeatism, it is rather a sense of tragedy in a world in which necessity is represented by an inevitable disappear-ance of differentiation. The declaration of our own nature and the attempt to build an enclave of organization in the face of nature's overwhelming tendency to disorder is an insolence against the gods and the iron necessity that they impose. Here lies tragedy, but here lies glory too.

Even when we discount the romantic, heroic overtones in that statement, Wiener is articulating what, as he saw and experienced it, makes living meaningful. The adjective 'arbitrary' before 'order and system' helps to make the statement appropriate for many; it might have been made by an artist as readily as by a creative scientist. Wiener's outlook on life is couched in the language of conflict and heroic struggle against overwhelming natural tendencies. But he was talking about something very different from the ruthless exploitation, even destruction, of nature and successfully bending it to human purposes, which is part of the legacy, part of the nineteenth-century heroic ideal, of Western man. Wiener in his discussion of human purposes, recognizing feedbacks and larger systems which include the environment, had moved far away from that ideal and closer to an ideal of understanding and, both consciously and effectively, of collaborating with natural processes.

I expect that Wiener would have welcomed some more recent developments in physics, as his thinking was already at times tending in that direction. Since his day developments in the field of statistical mechanics have come to modify the ideas about how orderly patterns – for example, the growth of plants and animals and the evolution of ecosystems – arise in the face of the second law of thermodynamics. As Wiener anticipated, the notions of information, feedback and non-linearity of the differential equations have become increasingly important in biology.

But beyond that, Ilya Prigogine and his co-workers in Belgium have more recently made a convincing case that natural systems which are either far from thermodynamic equilibrium initially, or which fluctuate, may not return to equilibrium at all (G. Nicolis and I. Prigogine, *Self-Organization in Nonequilibrium Systems*, 1977). Instead they continue to move still further away from equilibrium towards a different, increasingly complex and orderly, but nevertheless stable pattern – not necessarily static, but possibly cyclic. According to the American physicist Willard Gibbs' way of thinking, the stable state of a system – equilibrium – is independent of its detailed initial conditions, yet that simplification no longer holds for systems finding stability far from equilibrium. This is an explicit mechanism quite different from that of a 'Maxwell demon' (explained in

Chapter 2), the mechanism assumed necessary in Wiener's day. It is more nearly related to Wiener's notion of positive feedback, which he tended to see as only disruptive and destructive, rather than as leading to complex stable structures. The results obtained by the Prigogine group show the creation of orderly patterns – natural countertrends to the tendency towards disorganization – to be stronger and more ordinary and commonplace than a sole reliance on mechanisms of the Maxwell-demon type would suggest. Sensitivity to initial conditions is also a prominent feature of 'chaos theory', currently an active field of research.

If, however, we now extend Wiener's analogy from statistical mechanics and incorporate the findings of the Prigogine group – according to which natural and spontaneous mechanisms other than just the Maxwell demon generate organization and differentiation – this suggests a shift in emphasis from 'the human fight against the increase of entropy to create local enclaves of order' to a more co-operative endeavour which to a considerable extent occurs naturally and of its own accord. It is a subtle shift that can, however, make large differences. Yet to be explored, these differences appear to echo disagreements that some modern feminists, neo-Taoists and ecologists have with classical Greek concepts of the heroic and the tragic.

Wiener's status, which he strongly prized, was that of an independent scientifically knowledgeable intellectual. He avoided accepting funds from government agencies or corporations that might in any way compromise his complete honesty and independence. Nor did he identify himself with any political, social or philosophical group, but spoke and wrote simply as an individual. He was suspicious of honours and prizes given for scientific achievement. After receiving the accolade of election to the National Academy of Sciences, he resigned, lest membership in that select, exclusive body of scientists corrupt his autonomous status as outsider *vis-à-vis* the American scientific establishment. He was of the tradition in which it is the intellectual's responsibility to speak truth to power. This was in the post-war years, when the US government and many scientists and science administrators were celebrating the continuing partnership between government and science, government providing the funds and scientists engaging in research. Wiener remained aloof and highly critical of that peacetime arrangement. More precisely, he tried to stay aloof, but he would not separate himself completely because for many years he remained a professor at the Massachusetts Institute of Technology, an institution heavily involved in that partnership. As was his nature, he continued to talk to colleagues about his own fertile ideas, whether they dealt with mathematics, engineering or social concerns.

The Human Use of Human Beings, first published in 1950, was a sequel to an earlier volume, *Cybernetics: Or Control and Communication in the Animal and the Machine*. That earlier volume broke new ground in several respects. First of all, it was a report on new scientific and technical developments of the 1940s, especially on information theory, communication theory and communications technology, models of the brain and general-purpose computers. Secondly, it extended ideas and used metaphors from physics and electrical engineering to discuss a variety of topics including neuropathology, politics, society, learning and the nature of time.

Wiener had been an active participant in pre-war interdisciplinary seminars. After the war he regularly took part in a series of small conferences of mathematicians and engineers, which were also attended by biologists, anthropologists, sociologists, psychologists and psychiatrists, in which the set of ideas subsumed under cybernetics was explored in the light of these various disciplines. At these conferences Wiener availed himself of the convenient opportunity to become acquainted with current research on a broad range of topics outside of his speciality.

Already in his *Cybernetics* Wiener had raised questions about the benefits of the new ideas and technologies, concluding pessimistically,

there are those who hope that the good of a better understanding of man and society which is offered by this new field of work may anticipate and outweigh the incidental contribution we are making to the concentration of power. I write in 1947, and I am compelled to say that it is a very slight hope.

The book was a rarity also in that, along with the technical material, he discussed ethical issues at length. *The Human Use of Human Beings* is a popularization of *Cybernetics* (omitting the forbidding mathematics), though with a special emphasis on the description of the human and the social.

The present volume is a reprint of the second (1954) edition, which differs significantly from the original hard-cover edition. The notable reorganization of the book and the changes made deserve attention. In the first edition we read that 'the purpose of this book is both to explain the potentialities of the machine in fields which up to now have been taken to be purely human, and to warn against the dangers of a purely selfish exploitation of these possibilities in a world in which to human beings human things are all-important.' After commenting critically about patterns of social organization in which all orders come from above, and none return ('an ideal held by many Fascists, Strong Men in Business, and Government'), he explains, 'I wish to devote this book [first edition] to a protest against this inhuman use of human beings.' The second edition, in contrast, as stated in the Preface, is organized around Wiener's other major theme, 'the impact of the Gibbsian point of view on modern life, both through the substantive changes it has made in working science, and through the changes it has made indirectly in our attitude to life in general.' The second edition, where the framework is more philosophical and less political, appears to be presented in such a way as to make it of interest not only in 1954, but also for many years to come. The writing and the organization are a bit tighter and more orderly than in the first edition. It also includes comment on some exemplifications of cybernetics (e.g., the work of Ross Ashby) that had come to Wiener's attention only during the early 1950s. Yet, even though several chapters are essentially unchanged, something was lost in going from the first to the second edition. I miss the bluntness and pungency of some of the comments in the earlier edition, which apparently were 'cleaned up' for the second.

The *cause célèbre* in 1954 in the USA was the Oppenheimer case. J. Robert Oppenheimer, the physicist who had directed the building of atom bombs during World War II, had subsequently come to disagree with the politically dominant figures in the government who were eager to develop and build with the greatest possible speed hydrogen bombs a thousand times more powerful than the atom bombs which had devastated Hiroshima and Nagasaki. Oppenheimer urged delay, as he preferred that a further effort be made to negotiate with the Soviet Union before proceeding with such an irreversible escalation of the arms race. This policy difference lay behind the dramatic Oppenheimer hearings, humiliating proceedings at the height of the anti-Communist 'McCarthy era' (and of the US Congressional 'Un-American Activities Committee'), leading to, absurdly, the labelling of Oppenheimer as a 'security risk'.

In that political atmosphere it is not surprising for a publisher to prefer a different focus than the misuse of the latest technologies, or the dangers of capitalist exploitation of technologies for profit. Wiener himself was at that time going on a lecture tour to India and was then occupied with several other projects, such as writing the second volume of his autobiography, the mathematical analysis of brain waves, sensory prosthesis and a new formulation of quantum theory. He did not concern himself a great deal with the revision of a book he had written several years earlier – it would be more characteristic of him to write a new book or add a new chapter, rather than revise a book already written – although he must have agreed to all revisions and editorial changes.

At the end of the book, in both editions, Wiener compares the Catholic Church with the Communist Party, and both with cold war government activities in capitalist America. The criticisms of America in these last few pages of the first edition (see Appendix to this Introduction) are, in spite of one brief pointed reference to McCarthyism, largely absent in the second edition. There are other differences in the two editions. The chapter 'Progress and Entropy', for example, is much longer in the first edition. The section on the history of inventions within that chapter is more detailed. The chapter also deals with such topics as the depletion of resources and American dependence on other nations for oil, copper and tin, and the possibility of an energy-crisis unless new inventions obviate it. It reviews vividly the progress in medicine and anticipates new problems, such as the increasing use of synthetic foods that may contain minute quantities of carcinogens. These and other discursive excursions, peripheral to the main line of argument of the book, are omitted in the present edition.

The Human Use of Human Beings was not Wiener's last word on the subject. He continued to think and talk and write. In 1959 he addressed and provoked a gathering of scientists by his reflections and analysis of some moral and technical consequences of automation (*Science*, vol. 131, p. 1358, 1960), and in his last book (*God and Golem, Inc.*, 1964) he returned to ethical concerns from the perspective of the creative scientist or engineer.

It was Wiener's lifelong obsession to distinguish the human from the machine, having recognized the identity of patterns of organization and of many functions which can be performed by either, but in *The Human Use of Human Beings* it is his intention to place his understanding of the people/machines identity/dichotomy within the context of his generous and humane social philosophy. Cybernetics had originated from the analysis of formal analogies between the behaviour of organisms and that of electronic and mechanical systems. The mostly military technologies new in his day, which today we call 'artificial intelligence', highlighted the potential resemblance between certain elaborate machines and people. Academic psychology in North America was in those days still predominantly behaviourist. The cybernetic machines – such as general-purpose computers – suggested a possibility as to the nature of mind: mind was analogous to the formal structure and organization, or the software aspect, of a reasoning-and-perceiving machine that could also issue instructions leading to actions. Thus the long-standing mind–brain duality was overcome by a materialism which encompassed organization, messages and information in addition to stuff and matter. But the subjective – an individual's cumulative experience, sensations and feelings, including the subjective experience of being alive – is belittled, seen only within the context of evolutionary theory as providing information useful for survival to the organism.

If shorn of Wiener's benign social philosophy, what remains of cybernetics can be used within a highly mechanical and dehumanizing, even militaristic, outlook. The fact that the metaphor of a sophisticated automaton is so heavily employed invites thinking about humans as in effect machines. Many who have learned merely the technical aspects of cybernetics have used them, and do so today, for ends which Wiener abhorred. It is a danger he foresaw, would have liked to obviate and, although aware of how little he could do in that regard, valiantly tried to head off.

The technological developments in themselves are impressive, but since most of us already have to bear with a glut of promotional literature it is more to the point here to frame discussion not in the promoters' terms (what the new machine can do), but in a more human and social framework: how is the machine affecting people's lives? Or still more pointedly: who reaps a benefit from it? Wiener urged scientists and engineers to practise 'the imaginative forward glance' so as to attempt assessing the impact of an innovation, even before making it known.

However, once some of the machines or techniques were put on the market, a younger generation with sensitivity to human and social impacts could report empirically where the shoe pinches. Even though such reports may not suffice to radically change conventional patterns of deployment of technologies, which after all express many kinds of political and economic interests, they at least document what happens and help to educate the public. As long as their authors avoid an a priori pro-technology or anti-technology bias, they effectively carry on where Wiener left off. Among such reports we note Joseph Weizenbaum's description of the human damage manifested in the 'compulsive programmer', which poses questions about appropriate and inappropriate uses of computers (*Computer Power and Human Reason*, 1976). Similarly David Noble has documented how the introduction of automation in the machine-tool industry has resulted in a deskilling of machinists to their detriment, and has described in detail the political process by which this deskilling was brought about (*Forces of Production*, 1984).

These kinds of 'inhuman' uses seem nearly subtle if placed next to the potentially most damaging use, war. The growth of communication–computation–automation devices and systems had made relatively small beginnings during World War II, but since then has been given high priority in US government-subsidized military research and development, and in the Soviet Union as well; their proliferation in military contexts has been enormous and extensive. A proper critique would entail an analysis in depth of world politics, and especially the political relations of the two 'superpowers'. Wiener feared that he had helped to provide tools for the centralization of power, and indeed he and his fellow scientists and engineers had. For instance, under the Reagan government many billions of dollars were spent on plans for a protracted strategic nuclear war with the Soviet Union. The technological 'challenge' was seen to be the development of an effective C-cubed system (command, control and communication) which would be used to destroy enemy political and command centres and at the same time, through a multitude of methods, prevent the destruction of the corresponding American centres, leaving the USA fully in command throughout the nuclear war and victorious. Some principled scientists and engineers have, in a Wienerian spirit, refused to work on, or have stopped working on, such mad schemes, or on implementing the politicians' 'Star-Wars' fantasies.

We have already alluded to Wiener's heavy use of metaphors from engineering to describe the human and the social, and his neglect of the subjective experience. In the post-war years American sociologists, anthropologists, political scientists and psychologists tried harder than ever to be seen as 'scientific'. They readily borrowed the engineers' idiom and many sought to learn from the engineers' or mathematicians' thinking. Continental European social thinkers were far more inclined to attend to the human subject and to make less optimistic claims about their scientific expertise, but it required another decade before European thought substantially influenced the positivistic or logical-empiricist predilections of the mainstream of American social scientists.

A major development in academic psychology, prominent and well-funded today, relies strongly on the concept of information processing and models based on the computer. It traces its origins to the discussions on cybernetics in the post-war years and the wartime work of the British psychologist Kenneth Craik. This development, known as 'cognitive science', entirely ignores background contexts, the culture, the society, history, subjective experience, human feelings and emotions. Thus it works with a highly impoverished model of what it is to be human. Such models have, however, found their challengers and critics, ranging from the journalist Gordon Ratray Taylor (*The Natural History of Mind*, 1979) to the psychologist James J. Gibson, the latter providing a far different approach to how humans know and perceive (*The Perception of the Visual World*, 1950; *The Senses Considered as Perceptual Systems*, 1966; *The Ecological Approach to Visual Perception*, 1979).

If we trace the intellectual history of current thinking in such diverse fields as cellular biology, medicine, anthropology, psychiatry, ecology and economics, we find that in each discipline concepts coming from cybernetics consitute one of the streams that have fed it. Cybernetics, including information theory, systems with purposive behaviour and automaton models, was part of the intellectual dialogue of the 1950s and has since mingled with many other streams, has been absorbed and become part of the conventional idiom and practice.

Too many writings about technologies are dismal, narrow apologetics for special interests, and not very edifying. Yet the subject matter is intrinsically extremely varied and stimulating to an enquiring mind. It has profound implications for our day-to-day lives, their structure and their quality. The social history of science and technology is a rich resource, even for imagining and reflecting on the future. Moreover the topic highlights central dilemmas in every political system. For example, how is the role of 'experts' in advising governments related to political process? Or how is it possible to reconcile, in a capitalist economy within a democratic political structure, the unavoidable conflict between public interest and decision by a popular vote, on the one hand, and corporate decisions as to which engineering projects are profitable, on the other?

We are now seeing the rise of a relatively new genre of writing about technologies and people which is interesting, concrete, open, exploratory and confronts political issues head-on. We need this writing, for we are living in what Ellul has appropriately called a technological society. Within that genre, Wiener's books, as well as some earlier writings by Lewis Mumford, are among the few pioneering works that have become classics. The present reissue of one of these classics is cause for rejoicing. May it stimulate readers to think passionately for themselves about the human use of human beings with the kind of intellectual honesty and compassion Wiener brought to the subject.

Steve J. Heims
Boston, October 1988

APPENDIX

What follows are two documents from Norbert Wiener's writings:
– an open letter published in the *Atlantic Monthly* magazine, January 1947 issue; and
– the concluding passages of *The Human Use of Human Beings*, 1st edition, Houghton-Mifflin, 1950, pp. 228–9.

A SCIENTIST REBELS

The letter which follows was addressed by one of our ranking mathematicians to a research scientist of a great aircraft corporation, who had asked him for the technical account of a certain line of research he had conducted in the war. Professor Wiener's indignation at being requested to participate in indiscriminate rearmament, less than two years after victory, is typical of many American scientists who served their country faithfully during the war.

Professor of Mathematics in one of our great Eastern institutions, Norbert Wiener was born in Columbia, Missouri, in 1894, the son of Leo Wiener, Professor of Slavic Languages at Harvard University. He took his doctorate at Harvard and did his graduate work in England and in Göttingen. Today he is esteemed one of the world's foremost mathematical analysts. His ideas played a significant part in the development of the theories of communication and control which were essential in winning the war.
– The Editor, *Atlantic Monthly*

Sir:–

I have received from you a note in which you state that you are engaged in a project concerning controlled missiles, and in which you request a copy of a paper which I wrote for the National Defense Research Committee during the war.

As the paper is the property of a government organization, you are of course at complete liberty to turn to that government organization for such information as I could give you. If it is out of print as you say, and they desire to make it available for you, there are doubtless proper avenues of approach to them.

When, however, you turn to me for information concerning controlled missiles, there are several considerations which determine my reply. In the past, the comity of scholars has made it a custom to furnish scientific information to any person seriously seeking it. However, we must face these facts: the policy of the government itself during and after the war, say in the bombing of Hiroshima and Nagasaki, has made it clear that to provide scientific information is not a necessarily innocent act, and may entail the gravest consequences. One therefore cannot escape reconsidering the established custom of the scientist to give information to every person who may enquire of him. The interchange of ideas which is one of the great traditions of science must of course receive certain limitations when the scientist becomes an arbiter of life and death.

For the sake, however, of the scientist and the public, these limitations should be as intelligent as possible. The measures taken during the war by our military agencies, in restricting the free intercourse among scientists on related projects or even on the same project, have gone so far that it is clear that if continued in time of peace this policy will lead to the total irresponsibility of the scientist, and ultimately to the death of science. Both of these are disastrous for our civilization, and entail grave and immediate peril for the public.

I realize, of course, that I am acting as the censor of my own ideas, and it may sound arbitrary, but I will not accept a censorship in which I do not participate. The experience of the scientists who have worked on the atomic bomb has indicated that in any investigation of this kind the scientist ends by putting unlimited powers in the hands of the people whom he is least inclined to trust with their use. It is perfectly clear also that to disseminate information about a weapon in the present state of our civilization is to make it practically certain that that weapon will be used. In that respect the controlled missile represents the still imperfect supplement to the atom bomb and to bacterial warfare.

The practical use of guided missiles can only be to kill foreign civilians indiscriminately, and it furnishes no protection whatsoever to civilians in this country. I cannot conceive a situation in which such weapons can produce any effect other than extending the kamikaze way of fighting to whole nations. Their possession can do nothing but endanger us by encouraging the tragic insolence of the military mind.

If therefore I do not desire to participate in the bombing or poisoning of defenceless peoples – and I most certainly do not – I must take a serious responsibility as to those to whom I disclose my scientific ideas. Since it is obvious that with sufficient effort you can obtain my material, even though it is out of print, I can only protest *pro forma* in refusing to give you any information concerning my past work. However, I rejoice at the fact that my material is not readily available, inasmuch as it gives me the opportunity to raise this serious moral issue. I do not expect to publish any future work of mine which may do damage in the hands of irresponsible militarists.

I am taking the liberty of calling this letter to the attention of other people in scientific work. I believe it is only proper that they should know of it in order to make their own independent decisions, if similar situations should confront them.

Norbert Wiener

The Human Use of Human Beings

I have indicated that freedom of opinion at the present time is being crushed between the two rigidities of the Church and the Communist Party. In the United States we are in the process of developing a new rigidity which combines the methods of both while partaking of the emotional fervour of neither. Our Conservatives of all shades of opinion have somehow got together to make American capitalism and the fifth freedom of the businessman supreme throughout all the world.

Our military men and our great merchant princes have looked upon the propaganda technique of the Russians, and have found that it is good. They have found a worthy counterpart for the GPU in the FBI, in its new rôle of political censor. They have not considered that these weapons form something fundamentally distasteful to humanity, and that they need the full force of an overwhelming faith and belief to make them even tolerable. This faith and belief they have nowhere striven to replace. Thus they have been false to the dearest part of our American traditions, without offering us any principles for which we may die, except a merely negative hatred of Communism. They have succeeded in being un-American without being radical. To this end we have invented a new inquisition: the Inquisition of Teachers' Oaths and of Congressional Committees. We have synthesized a new propaganda, lacking only one element which is common to the Church and to the Communist Party, and that is the element of Belief. We have accepted the methods, not the ideals of our possible antagonists, little realizing that it is the ideals which have given the methods whatever cogency they possess. Ourselves without faith, we presume to punish heresy. May the absurdity of our position soon perish amidst the Homeric laughter that it deserves.

It is this triple attack on our liberties which we must resist, if communication is to have the scope that it properly deserves as the central phenomenon of society, and if the human individual is to reach and to maintain his full stature.

It is again the American worship of know-how as opposed to know-what that hampers us. We rightly see great dangers in the totalitarian system of Communism. On the one hand, we have called in to combat these the assistance of a totalitarian Church which is in no respect ready to accept, in support of its standards, milder means than those to which Communism appeals. On the other hand, we have attempted to synthesize a rigid system to fight fire by fire, and to oppose Communism by institutions which bear more than a fortuitous resemblance to Communistic institutions. In this we have failed to realize that the element in Communism which essentially deserves our respect consists in its loyalties and in its insistence on the dignity and the rights of the worker. What is bad consists chiefly in the ruthless techniques to which the present phase of the Communist revolution has resorted. Our leaders show a disquieting complacency in their acceptance of the ruthlessness and a disquieting unwillingness to refer their acts to any guiding principles. Fundamentally, behind our counter-ruthlessness there is no adequate basis of real heartfelt assent. Let us hope that it is still possible to reverse the tide of the moment and to create a future America in which man can live and can grow to be a human being in the fullest and richest sense of the word.

THE IDEA OF A CONTINGENT UNIVERSE

The beginning of the twentieth century marked more than the end of one hundred-year period and the start of another. There was a real change of point of view even before we made the political transition from the century on the whole dominated by peace, to the half century of war through which we have just been living. This was perhaps first apparent in science, although it is quite possible that whatever has affected science led independently to the marked break which we find between the arts and literature of the nineteenth and those of the twentieth centuries.

Newtonian physics, which had ruled from the end of the seventeenth century to the end of the nineteenth with scarcely an opposing voice, described a universe in which everything happened precisely according to law, a compact, tightly organized universe in which the whole future depends strictly upon the whole past. Such a picture can never be either fully justified or fully rejected experimentally and belongs in large measure to a conception of the world which is supplementary to experiment but in some ways more universal than anything that can be experimentally verified. We can never test by our imperfect experiments whether one set of physical laws or another can be verified down to the last decimal. The Newtonian view, however, was compelled to state and formulate physics as if it were, in fact, subject to such laws. This is now no longer the dominating attitude of physics, and the men who contributed most to its downfall were Bolzmann in Germany and Gibbs in the United States.

These two physicists undertook a radical application of an exciting, new idea. Perhaps the use of statistics in physics which, in large measure, they introduced was not completely new, for Maxwell and others had considered worlds of very large numbers of particles which necessarily had to be treated statistically. But what Bolzmann and Gibbs did was to introduce statistics into physics in a much more thoroughgoing way, so that the statistical approach was valid not merely for systems of enormous complexity, but even for systems as simple as the single particle in a field of force.

Statistics is the science of distribution, and the distribution contemplated by these modern scientists was not concerned with large numbers of similar particles, but with the various positions and velocities from which a physical system might start. In other words, under the Newtonian system the same physical laws apply to a variety of systems starting from a variety of positions and with a variety of momenta. The new statisticians put this point of view in a fresh light. They retained indeed the principle according to which certain systems may be distinguished from others by their total energy, but they rejected the supposition according to which systems with the same total energy may be clearly distinguished indefinitely and described forever by fixed causal laws.

There was, actually, an important statistical reservation implicit in Newton's work, though the eighteenth century, which lived by Newton, ignored it. No physical measurements are ever precise; and what we have to say about a machine or other dynamic system really concerns not what we must expect when the initial positions and momenta are given with perfect accuracy (which never occurs), but what we are to expect when they are given with attainable accuracy. This merely means that we know, not the complete initial conditions, but something about their distribution. The functional part of physics, in other words, cannot escape considering uncertainty and the contingency of events. It was the merit of Gibbs to show for the first time a clean-cut scientific method for taking this contingency into consideration.

The historian of science looks in vain for a single line of development. Gibbs' work, while well cut out, was badly sewed, and it remained for others to complete the job that he began. The intuition on which he based his work was that, in general, a physical system belonging to a class of physical systems, which continues to retain its identity as a class, eventually reproduces in almost all cases the distribution which it shows at any given time over the whole class of systems. In other words, under certain circumstances a system runs through all the distributions of position and momentum which are compatible with its energy, if it keeps running long enough.

This last proposition, however, is neither true nor possible in anything but trivial systems. Nevertheless, there is another route leading to the results which Gibbs needed to bolster his hypothesis. The irony of history is that this route was being explored very thoroughly in Paris at exactly the time when Gibbs was working in New Haven; and yet it was not until 1920 that the Paris work met the New Haven work in a fruitful union. I had, I believe, the honor of assisting at the birth of the first child of this union.

Gibbs had to work with theories of measure and probability which were already at least twenty-five years old and were grossly inadequate to his needs. At the same time, however, Borel and Lebesgue in Paris were devising the theory of integration which was to prove apposite to the Gibbsian ideas. Borel was a mathematician who had already made his reputation in the theory of probability and had an excellent physical sense. He did work leading to this theory of measure, but he did not reach the stage in which he could close it into a complete theory. This was done by his pupil Lebesgue, who was a very different sort of person. He had neither the sense of physics nor an interest in it. Nonetheless Lebesgue solved the problem put by Borel, but he regarded the solution of this problem as no more than a tool for Fourier series and other branches of pure mathematics. A quarrel developed between the two men when they both became candidates for admission to the French Academy of Sciences, and only after a great deal of mutual denigration, did they both receive this honor. Borel, however, continued to maintain the importance of Lebesgue's work and his own as a physical tool; but I believe that I myself, in 1920, was the first person to apply the Lebesgue integral to a specific physical problem—that of the Brownian motion.

This occurred long after Gibbs' death, and his work remained for two decades one of those mysteries of science which work even though it seems that they ought not to work. Many men have had intuitions well ahead of their time; and this is not least true in mathematical physics. Gibbs' introduction of probability into physics occurred well before there was an adequate theory of the sort of probability he needed. But for all these gaps it is, I am convinced, Gibbs rather than Einstein or Heisenberg or Planck to whom we must attribute the first great revolution of twentieth century physics.

This revolution has had the effect that physics now no longer claims to deal with what will always happen, but rather with what will happen with an overwhelming probability. At the beginning in Gibbs' own work this contingent attitude was superimposed on a Newtonian base in which the elements whose probability was to be discussed were systems obeying all of the Newtonian laws. Gibbs' theory was essentially new, but the permutations with which it was compatible were the same as those contemplated by Newton. What has happened to physics since is that the rigid Newtonian basis has been discarded or modified, and the Gibbsian contingency now stands in its complete nakedness as the full basis of physics. It is true that the books are not yet quite closed on this issue and that Einstein and, in some of his phases, De Broglie, still contend that a rigid deterministic world is more acceptable than a contingent one; but these great scientists are fighting a rear-guard action against the overwhelming force of a younger generation.

One interesting change that has taken place is that in a probabilistic world we no longer deal with quantities and statements which concern a specific, real universe as a whole but ask instead questions which may find their answers in a large number of similar universes. Thus chance has been admitted, not merely as a mathematical tool for physics, but as part of its warp and weft.

This recognition of an element of incomplete determinism, almost an irrationality in the world, is in a certain way parallel to Freud's admission of a deep irrational component in human conduct and thought. In the present world of political as well as intellectual confusion, there is a natural tendency to class Gibbs, Freud, and the proponents of the modern theory of probability together as representatives of a single tendency; yet I do not wish to press this point. The gap between the Gibbs-Lebesgue way of thinking and Freud's intuitive but somewhat discursive method is too large. Yet in their recognition of a fundamental element of chance in the texture of the universe itself, these men are close to one another and close to the tradition of St. Augustine. For this random element, this organic incompleteness, is one which without too violent a figure of speech we may consider evil; the negative evil which St. Augustine characterizes as incompleteness, rather than the positive malicious evil of the Manichaeans.

This book is devoted to the impact of the Gibbsian point of view on modern life, both through the substantive changes it has made in working science, and through the changes it has made indirectly in our attitude to life in general. Thus the following chapters contain an element of technical description as well as a philosophic component which concerns what we do and how we should react to the new world that confronts us.

I repeat: Gibbs' innovation was to consider not one world, but all the worlds which are possible answers to a limited set of questions concerning our environment. His central notion concerned the extent to which answers that we may give to questions about one set of worlds are probable among a larger set of worlds. Beyond this, Gibbs had a theory that this probability tended naturally to increase as the universe grows older. The measure of this probability is called entropy, and the characteristic tendency of entropy is to increase.

As entropy increases, the universe, and all closed systems in the universe, tend naturally to deteriorate and lose their distinctiveness, to move from the least to the most probable state, from a state of organization and differentiation in which distinctions and forms exist, to a state of chaos and sameness. In Gibbs' universe order is least probable, chaos most probable. But while the universe as a whole, if indeed there is a whole universe, tends to run down, there are local enclaves whose direction seems opposed to that of the universe at large and in which there is a limited and temporary tendency for organization to increase. Life finds its home in some of these enclaves. It is with this point of view at its core that the new science of Cybernetics began its development.[1]

[1] There are those who are skeptical as to the precise identity between entropy and biological disorganization. It will be necessary for me to evaluate these criticisms sooner or later, but for the present I must assume that the differences lie, not in the fundamental nature of these quantities, but in the systems in which they are observed. It is too much to expect a final, clear-cut definition of entropy on which all writers will agree in any less than the closed, isolated system.

THE HUMAN USE OF HUMAN BEINGS

I

CYBERNETICS IN HISTORY

Since the end of World War II, I have been working on the many ramifications of the theory of messages. Besides the electrical engineering theory of the transmission of messages, there is a larger field which includes not only the study of language but the study of messages as a means of controlling machinery and society, the development of computing machines and other such automata, certain reflections upon psychology and the nervous system, and a tentative new theory of scientific method. This larger theory of messages is a probabilistic theory, an intrinsic part of the movement that owes its origin to Willard Gibbs and which I have described in the introduction.

Until recently, there was no existing word for this complex of ideas, and in order to embrace the whole field by a single term, I felt constrained to invent one. Hence "Cybernetics," which I derived from the Greek word *kubernētēs*, or "steersman," the same Greek word from which we eventually derive our word "governor." Incidentally, I found later that the word had already been used by Ampère with reference to political science, and had been introduced in another context by a Polish scientist, both uses dating from the earlier part of the nineteenth century.

I wrote a more or less technical book entitled *Cybernetics* which was published in 1948. In response to a certain demand for me to make its ideas acceptable to the lay public, I published the first edition of *The Human Use of Human Beings* in 1950. Since then the subject has grown from a few ideas shared by Drs. Claude Shannon, Warren Weaver, and myself, into an established region of research. Therefore, I take this opportunity occasioned by the reprinting of my book to bring it up to date, and to remove certain defects and inconsequentialities in its original structure.

In giving the definition of Cybernetics in the original book, I classed communication and control together. Why did I do this? When I communicate with another person, I impart a message to him, and when he communicates back with me he returns a related message which contains information primarily accessible to him and not to me. When I control the actions of another person, I communicate a message to him, and although this message is in the imperative mood, the technique of communication does not differ from that of a message of fact. Furthermore, if my control is to be effective I must take cognizance of any messages from him which may indicate that the order is understood and has been obeyed.

It is the thesis of this book that society can only be understood through a study of the messages and the communication facilities which belong to it; and that in the future development of these messages and communication facilities, messages between man and machines, between machines and man, and between machine and machine, are destined to play an ever-increasing part.

When I give an order to a machine, the situation is not essentially different from that which arises when I give an order to a person. In other words, as far as my consciousness goes I am aware of the order that has gone out and of the signal of compliance that has come back. To me, personally, the fact that the signal in its intermediate stages has gone through a machine rather than through a person is irrelevant and does not in any case greatly change my relation to the signal. Thus the theory of control in engineering, whether human or animal or mechanical, is a chapter in the theory of messages.

Naturally there are detailed differences in messages and in problems of control, not only between a living organism and a machine, but within each narrower class of beings. It is the purpose of Cybernetics to develop a language and techniques that will enable us indeed to attack the problem of control and communication in general, but also to find the proper repertory of ideas and techniques to classify their particular manifestations under certain concepts.

The commands through which we exercise our control over our environment are a kind of information which we impart to it. Like any form of information, these commands are subject to disorganization in transit. They generally come through in less coherent fashion and certainly not more coherently than they were sent. In control and communication we are always fighting nature's tendency to degrade the organized and to destroy the meaningful; the tendency, as Gibbs has shown us, for entropy to increase.

Much of this book concerns the limits of communication within and among individuals. Man is immersed in a world which he perceives through his sense organs. Information that he receives is co-ordinated through his brain and nervous system until, after the proper process of storage, collation, and selection, it emerges through effector organs, generally his muscles. These in turn act on the external world, and also react on the central nervous system through receptor organs such as the end organs of kinaesthesia; and the information received by the kinaesthetic organs is combined with his already accumulated store of information to influence future action.

Information is a name for the content of what is exchanged with the outer world as we adjust to it, and make our adjustment felt upon it. The process of receiving and of using information is the process of our adjusting to the contingencies of the outer environment, and of our living effectively within that environment. The needs and the complexity of modern life make greater demands on this process of information than ever before, and our press, our museums, our scientific laboratories, our universities, our libraries and textbooks, are obliged to meet the needs of this process or fail in their purpose. To live effectively is to live with adequate information. Thus, communication and control belong to the essence of man's inner life, even as they belong to his life in society.

The place of the study of communication in the history of science is neither trivial, fortuitous, nor new. Even before Newton such problems were current in physics, especially in the work of Fermat, Huygens, and Leibnitz, each of whom shared an interest in physics whose focus was not mechanics but optics, the communication of visual images.

Fermat furthered the study of optics with his principle of minimization which says that over any sufficiently short part of its course, light follows the path which it takes the least time to traverse. Huygens developed the primitive form of what is now known as "Huygens' Principle" by saying that light spreads from a source by forming around that source something like a small sphere consisting of secondary sources which in turn propagate light just as the primary sources do. Leibnitz, in the meantime, saw the whole world as a collection of beings called "monads" whose activity consisted in the perception of one another on the basis of a pre-established harmony laid down by God, and it is fairly clear that he thought of this interaction largely in optical terms. Apart from this perception, the monads had no "windows," so that in his view all mechanical interaction really becomes nothing more than a subtle consequence of optical interaction.

A preoccupation with optics and with message, which is apparent in this part of Leibnitz's philosophy, runs through its whole texture. It plays a large part in two of his most original ideas: that of the *Characteristica Universalis,* or universal scientific language, and that of the *Calculus Ratiocinator,* or calculus of logic. This Calculus Ratiocinator, imperfect as it was, was the direct ancestor of modern mathematical logic.

Leibnitz, dominated by ideas of communication, is, in more than one way, the intellectual ancestor of the ideas of this book, for he was also interested in machine computation and in automata. My views in this book are very far from being Leibnitzian, but the problems with which I am concerned are most certainly Leibnitzian. Leibnitz's computing machines were only an offshoot of his interest in a computing language, a reasoning calculus which again was in his mind, merely an extention of his idea of a complete artificial language. Thus, even in his computing machine, Leibnitz's preoccupations were mostly linguistic and communicational.

Toward the middle of the last century, the work of Clerk Maxwell and of his precursor, Faraday, had attracted the attention of physicists once more to optics, the science of light, which was now regarded as a form of electricity that could be reduced to the mechanics of a curious, rigid, but invisible medium known as the ether, which, at the time, was supposed to permeate the atmosphere, interstellar space and all transparent materials. Clerk Maxwell's work on optics consisted in the mathematical development of ideas which had been previously expressed in a cogent but non-mathematical form by Faraday. The study of ether raised certain questions whose answers were obscure, as, for example, that of the motion of matter through the ether. The famous experiment of Michelson and Morley, in the nineties, was undertaken to resolve this problem, and it gave the entirely unexpected answer that there simply was no way to determine the motion of matter through the ether.

The first satisfactory solution to the problems aroused by this experiment was that of Lorentz, who pointed out that if the forces holding matter together were conceived as being themselves electrical or optical in nature, we should expect a negative result from the Michelson-Morley experiment. However, Einstein in 1905 translated these ideas of Lorentz into a form in which the unobservability of absolute motion was rather a postulate of physics than the result of any particular structure of matter. For our purposes, the important thing is that in Einstein's work, light and matter are on an equal basis, as they had been in the writings before Newton; without the Newtonian subordination of everything else to matter and mechanics.

In explaining his views, Einstein makes abundant use of the observer who may be at rest or may be moving. In his theory of relativity it is impossible to introduce the observer without also introducing the idea of message, and without, in fact, returning the emphasis of physics to a quasi-Leibnitzian state, whose tendency is once again optical. Einstein's theory of relativity and Gibbs' statistical mechanics are in sharp contrast, in that Einstein, like Newton, is still talking primarily in terms of an absolutely rigid dynamics not introducing the idea of probability. Gibbs' work, on the other hand, is probabilistic from the very start, yet both directions of work represent a shift in the point of view of physics in which the world as it actually exists is replaced in some sense or other by the world as it happens to be observed, and the old naïve realism of physics gives way to something on which Bishop Berkeley might have smiled with pleasure.

At this point it is appropriate for us to review certain notions pertaining to entropy which have already been presented in the introduction. As we have said, the idea of entropy represents several of the most important departures of Gibbsian mechanics from Newtonian mechanics. In Gibbs' view we have a physical quantity which belongs not to the outside world as such, but to certain sets of possible outside worlds, and therefore to the answer to certain specific questions which we can ask concerning the outside world. Physics now becomes not the discussion of an outside universe which may be regarded as the total answer to all the questions concerning it, but an account of the answers to much more limited questions. In fact, we are now no longer concerned with the study of all possible outgoing and incoming messages which we may send and receive, but with the theory of much more specific outgoing and incoming messages; and it involves a measurement of the no-longer infinite amount of information that they yield us.

Messages are themselves a form of pattern and organization. Indeed, it is possible to treat sets of messages as having an entropy like sets of states of the external world. Just as entropy is a measure of disorganization, the information carried by a set of messages is a measure of organization. In fact, it is possible to interpret the information carried by a message as essentially the negative of its entropy, and the negative logarithm of its probability. That is, the more probable the message, the less information it gives. Clichés, for example, are less illuminating than great poems.

I have already referred to Leibnitz's interest in automata, an interest incidentally shared by his contemporary, Pascal, who made real contributions to the development of what we now know as the desk adding-machine. Leibnitz saw in the concordance of the time given by clocks set at the same time, the model for the pre-established harmony of his monads. For the technique embodied in the automata of his time was that of the clockmaker. Let us consider the activity of the little figures which dance on the top of a music box. They move in accordance with a pattern, but it is a pattern which is set in advance, and in which the past activity of the figures has practically nothing to do with the pattern of their future activity. The probability that they will diverge from this pattern is nil. There is a message, indeed; but it goes from the machinery of the music box to the figures, and stops there. The figures themselves have no trace of communication with the outer world, except this one-way stage of communication with the pre-established mechanism of the music box. They are blind, deaf, and dumb, and cannot vary their activity in the least from the conventionalized pattern.

Contrast with them the behavior of man, or indeed of any moderately intelligent animal such as a kitten. I call to the kitten and it looks up. I have sent it a message which it has received by its sensory organs, and which it registers in action. The kitten is hungry and lets out a pitiful wail. This time it is the sender of a message. The kitten bats at a swinging spool. The spool swings to its left, and the kitten catches it with its left paw. This time messages of a very complicated nature are both sent and received within the kitten's own nervous system through certain nerve end-bodies in its joints, muscles, and tendons; and by means of nervous messages sent by these organs, the animal is aware of the actual position and tensions of its tissues. It is only through these organs that anything like a manual skill is possible.

I have contrasted the prearranged behavior of the little figures on the music box on the one hand, and the contingent behavior of human beings and animals on the other. But we must not suppose that the music box is typical of all machine behavior.

The older machines, and in particular the older attempts to produce automata, did in fact function on a closed clockwork basis. But modern automatic machines such as the controlled missile, the proximity fuse, the automatic door opener, the control apparatus for a chemical factory, and the rest of the modern armory of automatic machines which perform military or industrial functions, possess sense organs; that is, receptors for messages coming from the outside. These may be as simple as photoelectric cells which change electrically when a light falls on them, and which can tell light from dark, or as complicated as a television set. They may measure a tension by the change it produces in the conductivity of a wire exposed to it, or they may measure temperature by means of a thermocouple, which is an instrument consisting of two distinct metals in contact with one another through which a current flows when one of the points of contact is heated. Every instrument in the repertory of the scientific-instrument maker is a possible sense organ, and may be made to record its reading remotely through the intervention of appropriate electrical apparatus. Thus the machine which is conditioned by its relation to the external world, and by the things happening in the external world, is with us and has been with us for some time.

The machine which acts on the external world by means of messages is also familiar. The automatic photoelectric door opener is known to every person who has passed through the Pennsylvania Station in New York, and is used in many other buildings as well. When a message consisting of the interception of a beam of light is sent to the apparatus, this message actuates the door, and opens it so that the passenger may go through.

The steps between the actuation of a machine of this type by sense organs and its performance of a task may be as simple as in the case of the electric door; or it may be in fact of any desired degree of complexity within the limits of our engineering techniques. A complex action is one in which the data introduced, which we call the *input,* to obtain an effect on the outer world, which we call the *output,* may involve a large number of combinations. These are combinations, both of the data put in at the moment and of the records taken from the past stored data which we call the *memory*. These are recorded in the machine. The most complicated machines yet made which transform input data into output data are the high-speed electrical computing machines, of which I shall speak later in more detail. The determination of the mode of conduct of these machines is given through a special sort of input, which frequently consists of punched cards or tapes or of magnetized wires, and which determines the way in which the machine is going to act in one operation, as distinct from the way in which it might have acted in another. Because of the frequent use of punched or magnetic tape in the control, the data which are fed in, and which indicate the mode of operation of one of these machines for combining information, are called the *taping*.

I have said that man and the animal have a kinaesthetic sense, by which they keep a record of the position and tensions of their muscles. For any machine subject to a varied external environment to act effectively it is necessary that information concerning the results of its own action be furnished to it as part of the information on which it must continue to act. For example, if we are running an elevator, it is not enough to open the outside door because the orders we have given should make the elevator be at that door at the time we open it. It is important that the release for opening the door be dependent on the fact that the elevator is actually at the door; otherwise something might have detained it, and the passenger might step into the empty shaft. This control of a machine on the basis of its *actual* performance rather than its *expected* performance is known as *feedback*, and involves sensory members which are actuated by motor members and perform the function of *tell-tales* or *monitors*—that is, of elements which indicate a performance. It is the function of these mechanisms to control the mechanical tendency toward disorganization; in other words, to produce a temporary and local reversal of the normal direction of entropy.

I have just mentioned the elevator as an example of feedback. There are other cases where the importance of feedback is even more apparent. For example, a gun-pointer takes information from his instruments of observation, and conveys it to the gun, so that the latter will point in such a direction that the missile will pass through the moving target at a certain time. Now, the gun itself must be used under all conditions of weather. In some of these the grease is warm, and the gun swings easily and rapidly. Under other conditions the grease is frozen or mixed with sand, and the gun is slow to answer the orders given to it. If these orders are reinforced by an extra push given when the gun fails to respond easily to the orders and lags behind them, then the error of the gun-pointer will be decreased. To obtain a performance as uniform as possible, it is customary to put into the gun a control feedback element which reads the lag of the gun behind the position it should have according to the orders given it, and which uses this difference to give the gun an extra push.

It is true that precautions must be taken so that the push is not too hard, for if it is, the gun will swing past its proper position, and will have to be pulled back in a series of oscillations, which may well become wider and wider, and lead to a disastrous instability. If the feedback system is itself controlled—if, in other words, its own entropic tendencies are checked by still other controlling mechanisms—and kept within limits sufficiently stringent, this will not occur, and the existence of the feedback will increase the stability of performance of the gun. In other words, the performance will become less dependent on the frictional load; or what is the same thing, on the drag created by the stiffness of the grease.

Something very similar to this occurs in human action.

If I pick up my cigar, I do not will to move any specific muscles. Indeed in many cases, I do not know what those muscles are. What I do is to turn into action a certain feedback mechanism; namely, a reflex in which the amount by which I have yet failed to pick up the cigar is turned into a new and increased order to the lagging muscles, whichever they may be. In this way, a fairly uniform voluntary command will enable the same task to be performed from widely varying initial positions, and irrespective of the decrease of contraction due to fatigue of the muscles. Similarly, when I drive a car, I do not follow out a series of commands dependent simply on a mental image of the road and the task I am doing. If I find the car swerving too much to the right, that causes me to pull it to the left. This depends on the actual performance of the car, and not simply on the road; and it allows me to drive with nearly equal efficiency a light Austin or a heavy truck, without having formed separate habits for the driving of the two. I shall have more to say about this in the chapter in this book on special machines, where we shall discuss the service that can be done to neuropathology by the study of machines with defects in performance similar to those occurring in the human mechanism.

It is my thesis that the physical functioning of the living individual and the operation of some of the newer communication machines are precisely parallel in their analogous attempts to control entropy through feedback. Both of them have sensory receptors as one stage in their cycle of operation: that is, in both of them there exists a special apparatus for collecting information from the outer world at low energy levels, and for making it available in the operation of the individual or of the machine. In both cases these external messages are not taken *neat*, but through the internal transforming powers of the apparatus, whether it be alive or dead. The information is then turned into a new form available for the further stages of perform-ance. In both the animal and the machine this per-formance is made to be effective on the outer world. In both of them, their *performed* action on the outer world, and not merely their *intended* action, is re-ported back to the central regulatory apparatus. This complex of behavior is ignored by the average man, and in particular does not play the role that it should in our habitual analysis of society; for just as individ-ual physical responses may be seen from this point of view, so may the organic responses of society itself. I do not mean that the sociologist is unaware of the ex-istence and complex nature of communications in society, but until recently he has tended to overlook the extent to which they are the cement which binds its fabric together.

We have seen in this chapter the fundamental unity of a complex of ideas which until recently had not been sufficiently associated with one another, namely, the contingent view of physics that Gibbs introduced as a modification of the traditional, Newtonian con-ventions, the Augustinian attitude toward order and conduct which is demanded by this view, and the theory of the message among men, machines, and in society as a sequence of events in time which, though it itself has a certain contingency, strives to hold back nature's tendency toward disorder by adjusting its parts to various purposive ends.

II

PROGRESS AND ENTROPY

As we have said, nature's statistical tendency to disorder, the tendency for entropy to increase in isolated systems, is expressed by the second law of thermodynamics. We, as human beings, are not isolated systems. We take in food, which generates energy, from the outside, and are, as a result, parts of that larger world which contains those sources of our vitality. But even more important is the fact that we take in information through our sense organs, and we act on information received.

Now the physicist is already familiar with the significance of this statement as far as it concerns our relations with the environment. A brilliant expression of the role of information in this respect is provided by Clerk Maxwell, in the form of the so-called "Maxwell demon," which we may describe as follows.

Suppose that we have a container of gas, whose temperature is everywhere the same. Some molecules of this gas will be moving faster than others. Now let us suppose that there is a little door in the container that lets the gas into a tube which runs to a heat engine, and that the exhaust of this heat engine is connected by another tube back to the gas chamber, through another door. At each door there is a little being with the power of watching the on-coming molecules and of opening or closing the doors in accordance with their velocity.

The demon at the first door opens it only for high-speed molecules and closes it in the face of low-speed molecules coming from the container. The role of the demon at the second door is exactly the opposite: he opens the door only for low-speed molecules coming from the container and closes it in the face of high-speed molecules. The result is that the temperature goes up at one end and down at the other thus creating a perpetual motion of "the second kind": that is, a perpetual motion which does not violate the first law of thermodynamics, which tells us that the amount of energy within a given system is constant, but does violate the second law of thermodynamics, which tells us that energy spontaneously runs down hill in temperature. In other words, the Maxwell demon seems to overcome the tendency of entropy to increase.

Perhaps I can illustrate this idea still further by considering a crowd milling around in a subway at two turnstiles, one of which will only let people out if they are observed to be running at a certain speed, and the other of which will only let people out if they are moving slowly. The fortuitous movement of the people in the subway will show itself as a stream of fast-moving people coming from the first turnstile, whereas the second turnstile will only let through slow-moving people. If these two turnstiles are connected by a passageway with a treadmill in it, the fast-moving people will have a greater tendency to turn the treadmill in one direction than the slow people to turn it in the other, and we shall gather a source of useful energy in the fortuitous milling around of the crowd.

Here there emerges a very interesting distinction between the physics of our grandfathers and that of the present day. In nineteenth century physics, it seemed to cost nothing to get information. The result is that there is nothing in Maxwell's physics to prevent one of his demons from furnishing its own power source. Modern physics, however, recognizes that the demon can only gain the information with which it opens or closes the door from something like a sense organ which for these purposes is an eye. The light that strikes the demon's eye is not an energy-less supplement of mechanical motion, but shares in the main properties of mechanical motion itself. Light cannot be received by any instrument unless it hits it, and cannot indicate the position of any particle unless it hits the particle as well. This means, then, that even from a purely mechanical point of view we cannot consider the gas chamber as containing mere gas, but rather gas and light which may or may not be in equilibrium. If the gas and the light are in equilibrium, it can be shown as a consequence of present physical doctrine that the Maxwell demon will be as blind as if there were no light at all. We shall have a cloud of light coming from every direction, giving no indication of the position and momenta of the gas particles. Therefore the Maxwell demon will work only in a system that is not in equilibrium. In such a system, however, it will turn out that the constant collision between light and gas particles tends to bring the light and particles to an equilibrium. Thus while the demon may temporarily reverse the usual direction of entropy, ultimately it too will wear down.

The Maxwell demon can work indefinitely only if additional light comes from outside the system and does not correspond in temperature to the mechanical temperature of the particles themselves. This is a situation which should be perfectly familiar to us, because we see the universe around us reflecting light from the sun, which is very far from being in equilibrium with mechanical systems on the earth. Strictly speaking, we are confronting particles whose temperature is 50 or 60° F. with a light which comes from a sun at many thousands of degrees.

In a system which is not in equilibrium, or in part of such a system, entropy need not increase. It may, in fact, decrease locally. Perhaps this non-equilibrium of the world about us is merely a stage in a downhill course which will ultimately lead to equilibrium. Sooner or later we shall die, and it is highly probable that the whole universe around us will die the heat death, in which the world shall be reduced to one vast temperature equilibrium in which nothing really new ever happens. There will be nothing left but a drab uniformity out of which we can expect only minor and insignificant local fluctuations.

But we are not yet spectators at the last stages of the world's death. In fact these last stages can have no spectators. Therefore, in the world with which we are immediately concerned there are stages which, though they occupy an insignificant fraction of eternity, are of great significance for our purposes, for in them entropy does not increase and organization and its correlative, information, are being built up.

What I have said about these enclaves of increasing organization is not confined merely to organization as exhibited by living beings. Machines also contribute to a local and temporary building up of information, notwithstanding their crude and imperfect organization compared with that of ourselves.

Here I want to interject the semantic point that such words as life, purpose, and soul are grossly inadequate to precise scientific thinking. These terms have gained their significance through our recognition of the unity of a certain group of phenomena, and do not in fact furnish us with any adequate basis to characterize this unity. Whenever we find a new phenomenon which partakes to some degree of the nature of those which we have already termed "living phenomena," but does not conform to all the associated aspects which define the term "life," we are faced with the problem whether to enlarge the word "life" so as to include them, or to define it in a more restrictive way so as to exclude them. We have encountered this problem in the past in considering viruses, which show some of the tendencies of life—to persist, to multiply, and to organize—

but do not express these tendencies in a fully-developed form. Now that certain analogies of behavior are being observed between the machine and the living organism, the problem as to whether the machine is alive or not is, for our purposes, semantic and we are at liberty to answer it one way or the other as best suits our convenience. As Humpty Dumpty says about some of his more remarkable words, "I pay them extra, and make them do what I want."

If we wish to use the word "life" to cover all phenomena which locally swim upstream against the current of increasing entropy, we are at liberty to do so. However, we shall then include many astronomical phenomena which have only the shadiest resemblance to life as we ordinarily know it. It is in my opinion, therefore, best to avoid all question-begging epithets such as "life," "soul," "vitalism," and the like, and say merely in connection with machines that there is no reason why they may not resemble human beings in representing pockets of decreasing entropy in a framework in which the large entropy tends to increase.

When I compare the living organism with such a machine, I do not for a moment mean that the specific physical, chemical, and spiritual processes of life as we ordinarily know it are the same as those of life-imitating machines. I mean simply that they both can exemplify locally anti-entropic processes, which perhaps may also be exemplified in many other ways which we should naturally term neither biological nor mechanical.

While it is impossible to make any universal statements concerning life-imitating automata in a field which is growing as rapidly as that of automatization, there are some general features of these machines as they actually exist that I should like to emphasize. One is that they are machines to perform some definite task or tasks, and therefore must possess effector organs (analogous to arms and legs in human beings) with which such tasks can be performed. The second point is that they must be *en rapport* with the outer world by sense organs, such as photoelectric cells and thermometers, which not only tell them what the existing circumstances are, but enable them to record the performance or nonperformance of their own tasks. This last function, as we have seen, is called *feedback,* the property of being able to adjust future conduct by past performance. Feedback may be as simple as that of the common reflex, or it may be a higher order feedback, in which past experience is used not only to regulate specific movements, but also whole policies of behavior. Such a policy-feedback may, and often does, appear to be what we know under one aspect as a conditioned reflex, and under another as learning.

For all these forms of behavior, and particularly for the more complicated ones, we must have central decision organs which determine what the machine is to do next on the basis of information fed back to it, which it stores by means analogous to the memory of a living organism.

It is easy to make a simple machine which will run toward the light or run away from it, and if such machines also contain lights of their own, a number of them together will show complicated forms of social behavior such as have been described by Dr. Grey Walter in his book, *The Living Brain.* At present the more complicated machines of this type are nothing but scientific toys for the exploration of the possibilities of the machine itself and of its analogue, the nervous system. But there is reason to anticipate that the developing technology of the near future will use some of these potentialities.

Thus the nervous system and the automatic machine are fundamentally alike in that they are devices which make decisions on the basis of decisions they have made in the past. The simplest mechanical devices will make decisions between two alternatives, such as the closing or opening of a switch. In the nervous system, the individual nerve fiber also decides between carrying an impulse or not. In both the machine and the nerve, there is a specific apparatus for making future decisions depend on past decisions, and in the nervous system a large part of this task is done at those extremely complicated points called "synapses" where a number of incoming nerve fibers connect with a single outgoing nerve fiber. In many cases it is possible to state the basis of these decisions as a threshold of action of the synapse, or in other words, by telling how many incoming fibers should fire in order that the outgoing fibers may fire.

This is the basis of at least part of the analogy between machines and living organisms. The synapse in the living organism corresponds to the switching device in the machine. For further development of the detailed relationship between machines and living organisms, one should consult the extremely inspiring books of Dr. Walter and Dr. W. Ross Ashby.[1]

The machine, like the living organism, is, as I have said, a device which locally and temporarily seems to resist the general tendency for the increase of entropy. By its ability to make decisions it can produce around it a local zone of organization in a world whose general tendency is to run down.

The scientist is always working to discover the order and organization of the universe, and is thus playing a game against the arch enemy, disorganization. Is this devil Manichaean or Augustinian? Is it a contrary force opposed to order or is it the very absence of order itself? The difference between these two sorts of demons will make itself apparent in the tactics to be used against them. The Manichaean devil is an opponent,

[1] W. Ross Ashby, *Design for a Brain*, Wiley, New York, 1952, and W. Grey Walter, *The Living Brain*, Norton, New York, 1953.

like any other opponent, who is determined on victory and will use any trick of craftiness or dissimulation to obtain this victory. In particular, he will keep his policy of confusion secret, and if we show any signs of beginning to discover his policy, he will change it in order to keep us in the dark. On the other hand, the Augustinian devil, which is not a power in itself, but the measure of our own weakness, may require our full resources to uncover, but when we have uncovered it, we have in a certain sense exorcised it, and it will not alter its policy on a matter already decided with the mere intention of confounding us further. The Manichaean devil is playing a game of poker against us and will resort readily to bluffing; which, as von Neumann explains in his *Theory of Games,* is intended not merely to enable us to win on a bluff, but to prevent the other side from winning on the basis of a certainty that we will not bluff.

Compared to this Manichaean being of refined malice, the Augustinian devil is stupid. He plays a difficult game, but he may be defeated by our intelligence as thoroughly as by a sprinkle of holy water.

As to the nature of the devil, we have an aphorism of Einstein's which is more than an aphorism, and is really a statement concerning the foundations of scientific method. "The Lord is subtle, but he isn't simply mean." Here the word "Lord" is used to describe those forces in nature which include what we have attributed to his very humble servant, the Devil, and Einstein means to say that these forces do not bluff. Perhaps this devil is not far in meaning from Mephistopheles. When Faust asked Mephistopheles what he was, Mephistopheles replied, "A part of that force which always seeks evil and always does good." In other words, the devil is not unlimited in his ability to deceive, and the scientist who looks for a positive force determined to confuse us in the universe which he is investigating is wasting his time. Nature offers resistance to decoding, but it does not show ingenuity in finding new and undecipherable methods for jamming our communication with the outer world.

This distinction between the passive resistance of nature and the active resistance of an opponent suggests a distinction between the research scientist and the warrior or the game player. The research physicist has all the time in the world to carry out his experiments, and he need not fear that nature will in time discover his tricks and method and change her policy. Therefore, his work is governed by his best moments, whereas a chess player cannot make one mistake without finding an alert adversary ready to take advantage of it and to defeat him. Thus the chess player is governed more by his worst moments than by his best moments. I may be prejudiced about this claim: for I have found it possible myself to do effective work in science, while my chess has been continually vitiated by my carelessness at critical instants.

The scientist is thus disposed to regard his opponent as an honorable enemy. This attitude is necessary for his effectiveness as a scientist, but tends to make him the dupe of unprincipled people in war and in politics. It also has the effect of making it hard for the general public to understand him, for the general public is much more concerned with personal antagonists than with nature as an antagonist.

We are immersed in a life in which the world as a whole obeys the second law of thermodynamics: confusion increases and order decreases. Yet, as we have seen, the second law of thermodynamics, while it may be a valid statement about the whole of a closed system, is definitely not valid concerning a non-isolated part of it. There are local and temporary islands of decreasing entropy in a world in which the entropy as a whole tends to increase, and the existence of these islands enables some of us to assert the existence of progress. What can we say about the general direction of the battle between progress and increasing entropy in the world immediately about us?

The Enlightenment, as we all know, fostered the idea of progress, even though there were among the men of the eighteenth century some who felt that this progress was subject to a law of diminishing returns, and that the Golden Age of society would not differ very much from what they saw about them. The crack in the fabric of the Enlightenment, marked by the French Revolution, was accompanied by doubts of progress elsewhere. Malthus, for example, sees the culture of his age about to sink into the slough of an uncontrolled increase in population, swallowing up all the gains so far made by humanity.

The line of intellectual descent from Malthus to Darwin is clear. Darwin's great innovation in the theory of evolution was that he conceived of it not as a Lamarckian spontaneous ascent from higher to higher and from better to better, but as a phenomenon in which living beings showed (a) a spontaneous tendency to develop in many directions, and (b) a tendency to follow the pattern of their ancestors. The combination of these two effects was to prune an over-lush developing nature and to deprive it of those organisms which were ill-adapted to their environment, by a process of "natural selection." The result of this pruning was to leave a residual pattern of forms of life more or less well adapted to their environment. This residual pattern, according to Darwin, assumes the appearance of universal purposiveness.

The concept of a residual pattern has come to the fore again in the work of Dr. W. Ross Ashby. He uses it to explain the concept of machines that learn. He points out that a machine of rather random and haphazard structure will have certain near-equilibrium positions, and certain positions far from equilibrium, and that the near-equilibrium patterns will by their very nature last for a long time, while the others will appear only temporarily. The result is that in Ashby's machine, as in Darwin's nature, we have the appearance of a purposefulness in a system which is not purposefully constructed simply because purposelessness is in its very nature transitory. Of course, in the long run, the great trivial purpose of maximum entropy will appear to be the most enduring of all. But in the intermediate stages an organism or a society of organisms will tend to dally longer in those modes of activity in which the different parts work together, according to a more or less meaningful pattern.

I believe that Ashby's brilliant idea of the unpurposeful random mechanism which seeks for its own purpose through a process of learning is not only one of the great philosophical contributions of the present day, but will lead to highly useful technical developments in the task of automatization. Not only can we build purpose into machines, but in an overwhelming majority of cases a machine designed to avoid certain pitfalls of breakdown will look for purposes which it can fulfill.

Darwin's influence on the idea of progress was not confined to the biological world, even in the nineteenth century. All philosophers and all sociologists draw their scientific ideas from the sources available at their time. Thus it is not surprising to find that Marx and his contemporary socialists accepted a Darwinian point of view in the matter of evolution and progress.

In physics, the idea of progress opposes that of entropy, although there is no absolute contradiction between the two. In the forms of physics directly dependent on the work of Newton, the information which contributes to progress and is directed against the increase of entropy may be carried by extremely small quantities of energy, or perhaps even by no energy at all. This view has been altered in the present century by the innovation in physics known as *quantum theory*.

Quantum theory has led, for our purposes, to a new

association of energy and information. A crude form of this association occurs in the theories of line noise in a telephone circuit or an amplifier. Such background noise may be shown to be unavoidable, as it depends on the discrete character of the electrons which carry the current; and yet it has a definite power of destroying information. The circuit therefore demands a certain amount of communication power in order that the message may not be swamped by its own energy. More fundamental than this example is the fact that light itself has an atomic structure, and that light of a given frequency is radiated in lumps which are known as light quanta, which have a determined energy dependent on that frequency. Thus there can be no radiation of less energy than a single light quantum. The transfer of information cannot take place without a certain expenditure of energy, so that there is no sharp boundary between energetic coupling and informational coupling. Nevertheless, for most practical purposes, a light quantum is a very small thing; and the amount of energy transfer which is necessary for an effective informational coupling is quite small. It follows that in considering such a local process as the growth of a tree or of a human being, which depends directly or indirectly on radiation from the sun, an enormous local decrease in entropy may be associated with quite a moderate energy transfer. This is one of the fundamental facts of biology; and in particular of the theory of photosynthesis, or of the chemical process by which a plant is enabled to use the sun's rays to form starch, and other complicated chemicals necessary for life, out of the water and the carbon dioxide of the air.

Thus the question of whether to interpret the second law of thermodynamics pessimistically or not depends on the importance we give to the universe at large, on the one hand, and to the islands of locally decreasing entropy which we find in it, on the other. Remember that we ourselves constitute such an island of decreasing entropy, and that we live among other such islands. The result is that the normal prospective difference between the near and the remote leads us to give far greater importance to the regions of decreasing entropy and increasing order than to the universe at large. For example, it may very well be that life is a rare phenomenon in the universe; confined perhaps to the solar system, or even, if we consider life on any level comparable to that in which we are principally interested, to the earth alone. Nevertheless, we live on this earth, and the possible absence of life elsewhere in the universe is of no great concern to us, and certainly of no concern proportionate to the overwhelming size of the remainder of the universe.

Again, it is quite conceivable that life belongs to a limited stretch of time; that before the earliest geological ages it did not exist, and that the time may well come when the earth is again a lifeless, burnt-out, or frozen planet. To those of us who are aware of the extremely limited range of physical conditions under which the chemical reactions necessary to life as we know it can take place, it is a foregone conclusion that the lucky accident which permits the continuation of life in any form on this earth, even without restricting life to something like human life, is bound to come to a complete and disastrous end. Yet we may succeed in framing our values so that this temporary accident of living existence, and this much more temporary accident of human existence, may be taken as all-important positive values, notwithstanding their fugitive character.

In a very real sense we are shipwrecked passengers on a doomed planet. Yet even in a shipwreck, human decencies and human values do not necessarily vanish, and we must make the most of them. We shall go down, but let it be in a manner to which we may look forward as worthy of our dignity.

Up to this point we have been talking of a pessimism which is much more the intellectual pessimism of the professional scientist than an emotional pessimism which touches the layman. We have already seen that the theory of entropy, and the considerations of the ultimate heat-death of the universe, need not have such profoundly depressing moral consequences as they seem to have at first glance. However, even this limited consideration of the future is foreign to the emotional euphoria of the average man, and particularly to that of the average American. The best we can hope for the role of progress in a universe running downhill as a whole is that the vision of our attempts to progress in the face of overwhelming necessity may have the purging terror of Greek tragedy. Yet we live in an age not over-receptive to tragedy.

The education of the average American child of the upper middle class is such as to guard him solicitously against the awareness of death and doom. He is brought up in an atmosphere of Santa Claus; and when he learns that Santa Claus is a myth, he cries bitterly. Indeed, he never fully accepts the removal of this deity from his Pantheon, and spends much of his later life in the search for some emotional substitute.

The fact of individual death, the imminence of calamity, are forced upon him by the experiences of his later years. Nevertheless, he tries to relegate these unfortunate realities to the role of accidents, and to build up a Heaven on Earth in which unpleasantness has no place. This Heaven on Earth consists for him in an eternal progress, and a continual ascent to Bigger and Better Things.

Our worship of progress may be discussed from two points of view: a factual one and an ethical one—that is, one which furnishes standards for approval and disapproval. Factually, it asserts that the earlier advance of geographical discovery, whose inception corresponds to the beginning of modern times, is to be continued into an indefinite period of invention, of the discovery of new techniques for controlling the human environment. This, the believers in progress say, will go on and on without any visible termination in a future not too remote for human contemplation. Those who uphold the idea of progress as an ethical principle regard this unlimited and quasi-spontaneous process of change as a *Good Thing,* and as the basis on which they guarantee to future generations a Heaven on Earth. It is possible to believe in progress as a fact without believing in progress as an ethical principle; but in the catechism of many Americans, the one goes with the other.

Most of us are too close to the idea of progress to take cognizance either of the fact that this belief belongs only to a small part of recorded history, or of the other fact, that it represents a sharp break with our own religious professions and traditions. Neither for the Catholic, the Protestant, nor for the Jew, is the world a *good* place in which an enduring happiness is to be expected. The church offers its pay for virtue, not in any coin which passes current among the Kings of the Earth, but as a promissory note on Heaven.

In essence, the Calvinist accepts this too, with the additional dark note that the Elect of God who shall pass the dire final examination of Judgment Day are few, and are to be selected by His arbitrary decree. To secure this, no virtues on earth, no moral righteousness, may be expected to be of the slightest avail. Many a good man will be damned. The blessedness which the Calvinists do not expect to find for themselves even in Heaven, they certainly do not await on earth.

The Hebrew prophets are far from cheerful in their evaluation of the future of mankind, or even of their chosen Israel; and the great morality play of Job, while it grants him a victory of the spirit, and while the Lord deigns to return to him his flocks and his servants and his wives, nevertheless gives no assurance that such a relatively happy outcome will take place except through the arbitrariness of God.

The Communist, like the believer in progress, looks for his Heaven on Earth, rather than as a personal reward to be drawn on in a post-earthly individual existence. Nevertheless, he believes that this Heaven on Earth will not come of itself without a struggle. He is just as skeptical of the Big Rock Candy Mountains of the Future as of the Pie in the Sky when you Die. Nor is Islam, whose very name means resignation to the will of God, any more receptive to the ideal of progress. Of Buddhism, with its hope for Nirvana and a release from the external Wheel of Circumstance, I need say nothing; it is inexorably opposed to the idea of progress, and this is equally true for all the kindred religions of India.

Besides the comfortable passive belief in progress, which many Americans shared at the end of the nineteenth century, there is another one which seems to have a more masculine, vigorous connotation. To the average American, progress means the winning of the West. It means the economic anarchy of the frontier, and the vigorous prose of Owen Wister and Theodore Roosevelt. Historically the frontier is, of course, a perfectly genuine phenomenon. For many years, the development of the United States took place against the background of the empty land that always lay further to the West. Nevertheless, many of those who have waxed poetic concerning this frontier have been praisers of the past. Already in 1890, the census takes cognizance of the end of the true frontier conditions. The geographical limits of the great backlog of unconsumed and unbespoken resources of the country had clearly been set.

It is difficult for the average person to achieve an historical perspective in which progress shall have been reduced to its proper dimensions. The musket with which most of the Civil War was fought was

only a slight improvement over that carried at Water-
loo, and that in turn was nearly interchangeable with
the Brown Bess of Marlborough's army in the Low
Countries. Nevertheless, hand firearms had existed
since the fifteenth century or earlier, and cannon more
than a hundred years earlier still. It is doubtful
whether the smoothbore musket ever much exceeded
in range the best of the longbows, and it is certain that
it never equaled them in accuracy nor in speed of fire;
yet the longbow is the almost unimproved invention
of the Stone Age.

Again, while the art of shipbuilding had by no means
been completely stagnant, the wooden man-of-war,
just before it left the seas, was of a pattern which had
been fairly unchanged in its essentials since the early
seventeenth century, and which even then displayed
an ancestry going back many centuries more. One of
Columbus' sailors would have been a valuable able sea-
man aboard Farragut's ships. Even a sailor from the
ship that took Saint Paul to Malta would have been
quite reasonably at home as a forecastle hand on one of
Joseph Conrad's barks. A Roman cattleman from the
Dacian frontier would have made quite a competent
vaquero to drive longhorn steers from the plains of
Texas to the terminus of the railroad, although he
would have been struck with astonishment with what
he found when he got there. A Babylonian administra-
tor of a temple estate would have needed no training
either in bookkeeping or in the handling of slaves to
run an early Southern plantation. In short, the period
during which the main conditions of life for the vast
majority of men have been subject to repeated and
revolutionary changes had not even begun until the
Renaissance and the great voyages, and did not assume anything like the accelerated pace which we now
take for granted until well into the nineteenth century.

Under these circumstances, there is no use in looking
anywhere in earlier history for parallels to the success-

ful inventions of the steam engine, the steamboat, the locomotive, the modern smelting of metals, the telegraph, the transoceanic cable, the introduction of electric power, dynamite and the modern high explosive missile, the airplane, the electric valve, and the atomic bomb. The inventions in metallurgy which heralded the origin of the Bronze Age are neither so concentrated in time nor so manifold as to offer a good counter-example. It is very well for the classical economist to assure us suavely that these changes are purely changes in degree, and that changes in degree do not vitiate historic parallels. The difference between a medicinal dose of strychnine and a fatal one is also only one of degree.

Now, scientific history and scientific sociology are based on the notion that the various special cases treated have a sufficient similarity for the social mechanisms of one period to be relevant to those of another. However, it is certainly true that the whole scale of phenomena has changed sufficiently since the beginning of modern history to preclude any easy transfer to the present time of political, racial, and economic notions derived from earlier stages. What is almost as obvious is that the modern period beginning with the age of discovery is itself highly heterogeneous.

In the age of discovery Europe had become aware for the first time of the existence of great thinly-settled areas capable of taking up a population exceeding that of Europe itself; a land full of unexplored resources, not only of gold and silver but of the other commodities of commerce as well. These resources seemed inexhaustible, and indeed on the scale on which the society of 1500 moved, their exhaustion and the saturation of the population of the new countries were very remote. Four hundred and fifty years is farther than most people choose to look ahead.

However, the existence of the new lands encouraged an attitude not unlike that of Alice's Mad Tea Party.

When the tea and cakes were exhausted at one seat, the natural thing for the Mad Hatter and the March Hare was to move on and occupy the next seat. When Alice inquired what would happen when they came around to their original positions again, the March Hare changed the subject. To those whose past span of history was less than five thousand years and who were expecting that the Millennium and the Final Day of Judgment might overtake them in far less time, this Mad Hatter policy seemed most sensible. As time passed, the tea table of the Americas had proved not to be inexhaustible; and as a matter of fact, the rate at which one seat has been abandoned for the next has been increasing at what is probably a still increasing pace.

What many of us fail to realize is that the last four hundred years are a highly special period in the history of the world. The pace at which changes during these years have taken place is unexampled in earlier history, as is the very nature of these changes. This is partly the result of increased communication, but also of an increased mastery over nature which, on a limited planet like the earth, may prove in the long run to be an increased slavery to nature. For the more we get out of the world the less we leave, and in the long run we shall have to pay our debts at a time that may be very inconvenient for our own survival. We are the slaves of our technical improvement and we can no more return a New Hampshire farm to the self-contained state in which it was maintained in 1800 than we can, by taking thought, add a cubit to our stature or, what is more to the point, diminish it. We have modified our environment so radically that we must now modify ourselves in order to exist in this new environment. We can no longer live in the old one. Progress imposes not only new possibilities for the future but new restrictions. It seems almost as if progress itself and our fight against the increase of entropy intrinsically must end in the downhill path from which we are trying to escape. Yet this pessimistic sentiment is only conditional upon our blindness and inactivity, for I am convinced that once we become aware of the new needs that a new environment has imposed upon us, as well as the new means of meeting these needs that are at our disposal, it may be a long time yet before our civilization and our human race perish, though perish they will even as all of us are born to die. However, the prospect of a final death is far from a complete frustration of life and this is equally true for a civilization and for the human race as it is for any of its component individuals. May we have the courage to face the eventual doom of our civilization as we have the courage to face the certainty of our personal doom. The simple faith in progress is not a conviction belonging to strength, but one belonging to acquiescence and hence to weakness.

III

RIGIDITY AND LEARNING: TWO
PATTERNS OF COMMUNICATIVE BEHAVIOR

Certain kinds of machines and some living organisms —particularly the higher living organisms—can, as we have seen, modify their patterns of behavior on the basis of past experience so as to achieve specific anti-entropic ends. In these higher forms of communicative organisms the environment, considered as the past experience of the individual, can modify the pattern of behavior into one which in some sense or other will deal more effectively with the future environment. In other words, the organism is not like the clockwork monad of Leibnitz with its pre-established harmony with the universe, but actually seeks a new equilibrium with the universe and its future contingencies. Its present is unlike its past and its future unlike its present. In the living organism as in the universe itself, exact repetition is absolutely impossible.

The work of Dr. W. Ross Ashby is probably the greatest modern contribution to this subject insofar as it concerns the analogies between living organisms and machines. Learning, like more primitive forms of feedback, is a process which reads differently forward and backward in time. The whole conception of the apparently purposive organism, whether it is mechanical, biological, or social, is that of an arrow with a particular direction in the stream of time rather than that of a line segment facing both ways which we may regard as going in either direction. The creature that learns is not the mythical amphisbaena of the ancients, with a head at each end and no concern with where it is going. It moves ahead from a known past into an unknown future and this future is not interchangeable with that past.

Let me give still another example of feedback which will clarify its function with respect to learning. When the great control rooms at the locks of the Panama Canal are in use, they are two-way message centers. Not only do messages go out controlling the motion of the tow locomotives, the opening and closing of the sluices, and the opening and closing of the gates; but the control room is full of telltales which indicate not merely that the locomotives, the sluices, and the gates have received their orders, but that they have in fact effectively carried out these orders. If this were not the case, the lock master might very easily assume that a towing locomotive had stopped and might rush the huge mass of a battleship into the gates, or might cause any one of a number of similar catastrophes to take place.

This principle in control applies not merely to the Panama locks, but to states, armies, and individual human beings. When in the American Revolution, orders already drawn up had failed, through carelessness, to go from England commanding a British army to march down from Canada to meet another British army marching up from New York at Saratoga, Burgoyne's forces met a catastrophic defeat which a well conceived program of two-way communications would have avoided. It follows that administrative officials, whether of a government or a university or a corporation, should take part in a two-way stream of communication, and not merely in one descending from the top. Otherwise, the top officials may find that they have based their policy on a complete misconception of the facts that their underlings possess. Again, there is no task harder for a lecturer than to speak to a dead-pan audience. The purpose of applause in the theater—and it is essential—is to establish in the performer's mind some modicum of two-way communication.

This matter of social feedback is of very great sociological and anthropological interest. The patterns of communication in human societies vary widely. There are communities like the Eskimos, among whom there seems to be no chieftainship and very little subordination, so that the basis of the social community is simply the common desire to survive against enormous odds of climate and food supply. There are socially stratified communities such as are found in India, in which the means of communication between two individuals are closely restricted and modified by their ancestry and position. There are communities ruled by despots, in which every relation between two subjects becomes secondary to the relation between the subject and his king. There are the hierarchical feudal communities of lord and vassal, and the very special techniques of social communication which they involve.

Most of us in the United States prefer to live in a moderately loose social community, in which the blocks to communication among individuals and classes are not too great. I will not say that this ideal of communication is attained in the United States. Until white supremacy ceases to belong to the creed of a large part of the country it will be an ideal from which we fall short. Yet even this modified formless democracy is too anarchic for many of those who make efficiency their first ideal. These worshipers of efficiency would like to have each man move in a social orbit meted out to him from his childhood, and perform a function to which he is bound as the serf was bound to the clod. Within the American social picture, it is shameful to have these yearnings, and this denial of opportunities implied by an uncertain future. Accordingly, many of those who are most attached to this orderly state of permanently allotted functions would be confounded if they were forced to admit this publicly. They are only in a position to display their clear preferences through their actions. Yet these actions stand out distinctly enough. The businessman who separates himself from his employees by a shield of yes-men, or the head of a big laboratory who assigns each subordinate a particular problem, and begrudges him the privilege of thinking for himself so that he can move beyond his immediate problem and perceive its general relevance, show that the democracy to which they pay their respects is not really the order in which they would prefer to live. The regularly ordered state of pre-assigned functions toward which they gravitate is suggestive of the Leibnitzian automata and does not suggest the irreversible movement into a contingent future which is the true condition of human life.

In the ant community, each worker performs its proper functions. There may be a separate caste of soldiers. Certain highly specialized individuals perform the functions of king and queen. If man were to adopt this community as a pattern, he would live in a fascist state, in which ideally each individual is conditioned from birth for his proper occupation: in which rulers are perpetually rulers, soldiers perpetually soldiers, the peasant is never more than a peasant, and the worker is doomed to be a worker.

It is a thesis of this chapter that this aspiration of the fascist for a human state based on the model of the ant results from a profound misapprehension both of the nature of the ant and of the nature of man. I wish to point out that the very physical development of the insect conditions it to be an essentially stupid and unlearning individual, cast in a mold which cannot be modified to any great extent. I also wish to show how these physiological conditions make it into a cheap mass-produced article, of no more individual value than a paper pie plate to be thrown away after it is once used. On the other hand, I wish to show that the human individual, capable of vast learning and study, which may occupy almost half of his life, is physically equipped, as the ant is not, for this capacity. Variety and possibility are inherent in the human sensorium—and are indeed the key to man's most noble flights—because variety and possibility belong to the very structure of the human organism.

While it is possible to throw away this enormous advantage that we have over the ants, and to organize the fascist ant-state with human material, I certainly believe that this is a degradation of man's very nature, and economically a waste of the great human values which man possesses.

I am afraid that I am convinced that a community of human beings is a far more useful thing than a community of ants; and that if the human being is condemned and restricted to perform the same functions over and over again, he will not even be a good ant, not to mention a good human being. Those who would organize us according to permanent individual functions and permanent individual restrictions condemn the human race to move at much less than half-steam. They throw away nearly all our human possibilities and by limiting the modes in which we may adapt ourselves to future contingencies, they reduce our chances for a reasonably long existence on this earth.

Let us now turn to a discussion of the restrictions on the make-up of the ant which have turned the ant community into the very special thing it is. These restrictions have a deep-seated origin in the anatomy and the physiology of the individual insect. Both the insect and the man are air-breathing forms, and represent the end of a long transition from the easygoing life of the waterborne animal to the much more exacting demands of the land-bound. This transition from water to land, wherever it has occurred, has involved radical improvements in breathing, in the circulation generally, in the mechanical support of the organism, and in the sense organs.

The mechanical reinforcement of the bodies of land animals has taken place along several independent lines. In the case of most of the mollusks, as well as in the case of certain other groups which, though unrelated, have taken on a generally mollusk-like form, part of the outer surface secretes a non-living mass of calcareous tissue, the shell. This grows by accretion from an early stage in the animal until the end of its life. The spiral and helical forms of those groups need only this process of accretion to account for them.

If the shell is to remain an adequate protection for the animal, and the animal grows to any considerable size in its later stages, the shell must be a very appreciable burden, suitable only for land animals of the slowly moving and inactive life of the snail. In other shell-bearing animals, the shell is lighter and less of a load, but at the same time much less of a protection. The shell structure, with its heavy mechanical burden, has had only a limited success among land animals.

Man himself represents another direction of development—a direction found throughout the vertebrates, and at least indicated in invertebrates as highly developed as the limulus and the octopus. In all these forms, certain internal parts of the connective tissue assume a consistency which is no longer fibrous, but rather that of a very hard, stiff jelly. These parts of the body are called *cartilage,* and they serve to attach the powerful muscles which animals need for an active life. In the higher vertebrates, this primary cartilaginous skeleton serves as a temporary scaffolding for a skeleton of much harder material: namely, bone, which is even more satisfactory for the attachment of powerful muscles. These skeletons, of bone or cartilage, contain a great deal of tissue which is not in any strict sense alive, but throughout this mass of intercellular tissue there is a living structure of cells, cellular membranes, and nutritive blood vessels.

The vertebrates have developed not only internal

skeletons, but other features as well which suit them for active life. Their respiratory system, whether it takes the form of gills or lungs, is beautifully adapted to the active interchange of oxygen between the external medium and a blood, and the latter is made much more efficient than the average invertebrate blood by having its oxygen-carrying respiratory pigment concentrated in corpuscles. This blood is pumped through a closed system of vessels, rather than through an open system of irregular sinuses, by a heart of relatively high efficiency.

The insects and crustaceans, and in fact all the arthropods, are built for quite another kind of growth. The outer wall of the body is surrounded by a layer of chitin secreted by the cells of the epidermis. This chitin is a stiff substance rather closely related to cellulose. In the joints the layer of chitin is thin and moderately flexible, but over the rest of the animal it becomes that hard external skeleton which we see on the lobster and the cockroach. An internal skeleton such as man's can grow with the animal. An external skeleton (unless, like the shell of the snail, it grows by accretion) cannot. It is dead tissue, and possesses no intrinsic capability of growth. It serves to give a firm protection to the body and an attachment for the muscles, but it amounts to a strait jacket.

Internal growth among the arthropods can be converted into external growth only by discarding the old strait jacket, and by developing under it a new one, which is initially soft and pliable and can take a slightly new and larger form, but which very soon acquires the rigidity of its predecessor. In other words, the stages of growth are marked by definite moults, relatively frequent in the crustacean, and much less so in the insect. There are several such stages possible during the larval period. The pupal period represents a transition moult, in which the wings, that have not been functional in the larva, develop internally toward a functional condition. This becomes realized when the pre-final pupal stage, and the moult which terminates it gives rise to a perfect adult. The adult never moults again. It is in its sexual stage and although in most cases it remains capable of taking nourishment, there are insects in which the adult mouth-parts and the digestive tube are aborted, so that the *imago,* as it is called, can only mate, lay eggs, and die.

The nervous system takes part in this process of tearing down and building up. While there is a certain amount of evidence that some memory persists from the larva through to the imago, this memory cannot be very extensive. *The physiological condition for memory and hence for learning seems to be a certain continuity of organization, which allows the alterations produced by outer sense impressions to be retained as more or less permanent changes of structure or function.* Metamorphosis is too radical to leave much lasting record of these changes. It is indeed hard to conceive of a memory of any precision which can survive this process of radical internal reconstruction.

There is another limitation on the insect, which is due to its method of respiration and circulation. The heart of the insect is a very poor and weak tubular structure, which opens, not into well-defined blood vessels, but into vague cavities or sinuses conveying the blood to the tissues. This blood is without pigmented corpuscles, and carries the blood-pigments in solution. This mode of transferring oxygen seems to be definitely inferior to the corpuscular method.

In addition, the insect method of oxygenation of the tissues makes at most only local use of the blood. The body of the animal contains a system of branched tubules, carrying air directly from the outside into the tissues to be oxygenated. These tubules are stiffened against collapse by spiral fibers of chitin, and are thus passively open, but there is nowhere evidence of an active and effective system of air pumping. Respiration occurs by diffusion alone.

Notice that the same tubules carry by diffusion the good air in and the spent air, polluted with carbon dioxide, out to the surface. In a diffusion mechanism, the time of diffusion varies not as the length of the tube, but as the square of the length. Thus, in general, the efficiency of this system tends to fall off very rapidly with the size of the animal, and falls below the point of survival for an animal of any considerable size. So not only is the insect structurally incapable of a first-rate memory, he is also structurally incapable of an effective size.

To know the significance of this limitation in size, let us compare two artificial structures—the cottage and the skyscraper. The ventilation of a cottage is quite adequately taken care of by the leak of air around the window frames, not to mention the draft of the chimney. No special ventilation system is necessary. On the other hand, in a skyscraper with rooms within rooms, a shutdown of the system of forced ventilation would be followed in a very few minutes by an intolerable foulness of the air in the work spaces. Diffusion and even convection are no longer enough to ventilate such a structure.

The absolute maximum size of an insect is smaller than that attainable by a vertebrate. On the other hand, the ultimate elements of which the insect is composed are not always smaller than they are in man, or even in a whale. The nervous system partakes of this small size, and yet consists of neurons not much smaller than those in the human brain, though there are many fewer of them, and their structure is far less complex. In the matter of intelligence, we should expect that it is not only the relative size of the nervous system that counts, but in a large measure its absolute size. There is simply no room in the reduced structure of an insect for a nervous system of great complexity, nor for a large stored memory.

In view of the impossibility of a large stored memory, as well as of the fact that the youth of an insect such as an ant is spent in a form which is insulated from the adult phase by the intermediate catastrophe of metamorphosis, there is no opportunity for the ant to learn much. Add to this, that its behavior in the adult stage must be substantially perfect from the beginning, and it then becomes clear that the instructions received by the insect nervous system must be pretty much a result of the way it is built, and not of any personal experience. Thus the insect is rather like the kind of computing machine whose instructions are all set forth in advance on the "tapes," and which has next to no feedback mechanism to see it through the uncertain future. The behavior of an ant is much more a matter of instinct than of intelligence. *The physical strait jacket in which an insect grows up is directly responsible for the mental strait jacket which regulates its pattern of behavior.*

Here the reader may say: "Well, we already know that the ant as an individual is not very intelligent, so why all this fuss about explaining why it cannot be intelligent?" The answer is that *Cybernetics takes the view that the structure of the machine or of the organism is an index of the performance that may be expected from it.* The fact that the mechanical rigidity of the insect is such as to limit its intelligence while the mechanical fluidity of the human being provides for his almost indefinite intellectual expansion is highly relevant to the point of view of this book. Theoretically, if we could build a machine whose mechanical structure duplicated human physiology, then we could have a machine whose intellectual capacities would duplicate those of human beings.

In the matter of rigidity of behavior, the greatest contrast to the ant is not merely the mammal in general, but man in particular. It has frequently been observed that man is a neoteinic form: that is, that if we compare man with the great apes, his closest relatives, we find that mature man in hair, head, shape, body proportions, bony structure, muscles, and so on, is more like the newborn ape than the adult ape. Among the animals, man is a Peter Pan who never grows up.

This immaturity of anatomical structure corresponds to man's prolonged childhood. Physiologically, man does not reach puberty until he has already completed a fifth of his normal span of life. Let us compare this with the ratio in the case of a mouse, which lives three years and starts breeding at the end of three months. This is a ratio of twelve to one. The mouse's ratio is much more nearly typical of the large majority of mammals than is the human ratio.

Puberty for most mammals either represents the end of their epoch of tutelage, or is well beyond it. In our community, man is recognized as immature until the age of twenty-one, and the modern period of education for the higher walks of life continues until about thirty, actually beyond the time of greatest physical strength. Man thus spends what may amount to forty per cent of his normal life as a learner, again for reasons that have to do with his physical structure. It is as completely natural for a human society to be based on learning as for an ant society to be based on an inherited pattern.

Man like all other organisms lives in a contingent universe, but man's advantage over the rest of nature is that he has the physiological and hence the intellectual equipment to adapt himself to radical changes in his environment. The human species is strong only insofar as it takes advantage of the innate adaptive, learning faculties that its physiological structure makes possible.

We have already indicated that effective behavior must be informed by some sort of feedback process, telling it whether it has equalled its goal or fallen short. The simplest feedbacks deal with gross successes or failures of performance, such as whether we have actually succeeded in grasping an object that we have tried to pick up, or whether the advance guard of an army is at the appointed place at the appointed time. However, there are many other forms of feedback of a more subtle nature.

It is often necessary for us to know whether a whole policy of conduct, a strategy so to say, has proved successful or not. The animal we teach to traverse a maze in order to find food or to avoid electric shocks, must be able to record whether the general plan of running through the maze has been on the whole successful or not, and it must be able to change this plan in order to run the maze efficiently. This form of learning is most certainly a feedback, but it is a feedback on a higher level, a feedback of policies and not of simple actions. It differs from more elementary feedbacks in what Bertrand Russell would call its "logical type."

This pattern of behavior may also be found in machines. A recent innovation in the technique of telephonic switching provides an interesting mechanical analogy to man's adaptive faculty. Throughout the telephone industry, automatic switching is rapidly completing its victory over manual switching, and it may seem to us that the existing forms of automatic switching constitute a nearly perfect process. Nevertheless, a little thought will show that the present process is very wasteful of equipment. The number of people with whom I actually wish to talk over the telephone is limited, and in large measure is the same limited group day after day and week after week. I use most of the telephone equipment available to me to communicate with members of this group. Now, as the present technique of switching generally goes, the process of reaching one of the people whom we call up four or five times a day is in no way different from the process of reaching those people with whom we may never have a conversation. From the standpoint of balanced service, we are using either too little equipment to handle the frequent calls or too much to handle the infrequent calls, a situation which reminds me of Oliver Wendell Holmes' poem on the "one-hoss shay." This hoary vehicle, as you recollect, after one hundred years of service, showed itself to be so carefully designed that neither wheel, nor top, nor shafts, nor seat contained any part which manifested an uneconomical excess of wearing power over any other part. Actually, the "one-hoss shay" represents the pinnacle of engineering, and is not merely a humorous fantasy. If the tires had lasted a moment longer than the spokes or the dashboard than the shafts, these parts would have carried into disuse certain economic values. These values could either have been reduced without hurting the durability of the vehicle as a whole, or they could have been transferred equally throughout the entire vehicle to make the whole thing last longer. Indeed, any structure not of the nature of the "one-hoss shay" is wastefully designed.

This means that for the greatest economy of service it is not desirable that the process of my connection with Mr. A., whom I call up three times a day, and with Mr. B., who is for me only an unnoticed item in the telephone directory, should be of the same order. If I were allotted a slightly more direct means of connection with Mr. A., then the time wasted in having to wait twice as long for Mr. B. would be more than compensated for. If then, it is possible without excessive cost to devise an apparatus which will record my past conversations, and reapportion to me a degree of service corresponding to the frequency of my past use of the telephone channels, I should obtain a better service, or a less expensive one, or both. The Philips Lamp Company in Holland has succeeded in doing this. The quality of its service has been improved by means of a feedback of Russell's so-called "higher logical type." It is capable of greater variety, more adaptability, and deals more effectively than conventional equipment with the entropic tendency for the more probable to overwhelm the less probable.

I repeat, feedback is a method of controlling a system by reinserting into it the results of its past performance. If these results are merely used as numerical data for the criticism of the system and its regulation, we have the simple feedback of the control engineers. If, however, the information which proceeds backward from the performance is able to change the general method and pattern of performance, we have a process which may well be called learning.

Another example of the learning process appears in connection with the problem of the design of prediction machines. At the beginning of World War II, the comparative inefficiency of anti-aircraft fire made it necessary to introduce apparatus which would follow the position of an airplane, compute its distance, determine the length of time before a shell could reach it, and figure out where it would be at the end of that time. If the plane were able to take a perfectly arbitrary evasive action, no amount of skill would permit us to fill in the as yet unknown motion of the plane between the time when the gun was fired and the time when the shell should arrive approximately at its goal. However, under many circumstances the aviator either does not, or cannot, take arbitrary evasive action. He is limited by the fact that if he makes a rapid turn, centrifugal force will render him unconscious; and by the other fact that the control mechanism of his plane and the course of instructions which he has received practically force on him certain regular habits of control which show themselves even in his evasive action. These regularities are not absolute but are rather statistical preferences which appear most of the time. They may be different for different aviators, and they will certainly be for different planes. Let us remember that in the pursuit of a target as rapid as an airplane, there is not time for the computer to take out his instruments and figure where the plane is going to be. All the figuring must be built into the gun control itself. This figuring must include data which depend on our past statistical experience of airplanes of a given type under varying flight conditions. The present stage of anti-aircraft fire consists in an apparatus which uses either fixed data of this sort, or a selection among a limited number of such fixed data. The proper choice among these may be switched in by means of the voluntary action of the gunner.

However, there is another stage of the control problem which may also be dealt with mechanically. The problem of determining the flight statistics of a plane from the actual observation of its flight, and then of transforming these into rules for controlling the gun, is itself a definite and mathematical one. Compared with the actual pursuit of the plane, in accordance with given rules, it is a relatively slow action, and involves a considerable observation of the past flight of the airplane. It is nevertheless not impossible to mechanize this long-time action as well as the short-time action. We thus may construct an anti-aircraft gun which observes by itself the statistics concerning the motion of the target plane, which then works these into a system of control, and which finally adopts this system of control as a quick way for adjusting its position to the observed position and motion of the plane.

To my knowledge this has not yet been done, but it is a problem which lies along lines we are considering, and expect to use in other problems of prediction. The adjustment of the general plan of pointing and firing the gun according to the particular system of motions which the target has made is essentially an act of learning. It is a change in the *taping* of the gun's computing mechanism, which alters not so much the numerical data, as the process by which they are interpreted. It is, in fact, a very general sort of feedback, affecting the whole method of behavior of the instrument.

The advanced process of learning which we have here discussed is still limited by the mechanical conditions of the system in which it occurs, and clearly does not correspond to the normal process of learning in man. But from this process we can infer quite different ways in which learning of a complex sort can be mechanized. These indications are given respectively by the Lockean theory of association, and by Pavlov's theory of the conditioned reflex. Before I take these up, however, I wish to make some general remarks to cover in advance certain criticisms of the suggestion that I shall present.

Let me recount the basis on which it is possible to develop a theory of learning. By far the greater part of the work of the nerve physiologist has been on the conduction of impulses by nerve fibers or neurons, and this process is given as an all-or-none phenomenon. That is, if a stimulus reaches the point or threshold where it will travel along a nerve fiber at all, and not die out in a relatively short distance, the effect which it produces at a comparatively remote point on the nerve fiber is substantially independent of its initial strength.

These nerve impulses travel from fiber to fiber across connections known as *synapses*, in which one ingoing fiber may come in contact with many outgoing fibers, and one outgoing fiber in contact with many ingoing fibers. In these synapses, the impulse given by a single incoming nerve fiber is often not enough to produce an effective outgoing impulse. In general, if the impulses arriving at a given outgoing fiber by incoming synaptic connections are too few, the outgoing fiber will not respond. When I say too few, I do not necessarily mean that all incoming fibers act alike, nor even that with any set of incoming active synaptic connections the question of whether the outgoing fiber will respond may be settled once for all. I also do not intend to ignore the fact that some incoming fibers, instead of tending to produce a stimulus in the outgoing fibers with which they connect, may tend to prevent these fibers from accepting new stimuli.

Be that as it may, while the problem of the conduction of impulses along a fiber may be described in a rather simple way as an all-or-none phenomenon, the problem of the transmission of an impulse across a layer of synaptic connections depends on a complicated pattern of responses, in which certain combinations of incoming fibers, firing within a certain limited time, will cause the message to go further, while certain other combinations will not. These combinations are not a thing fixed once for all, nor do they even depend solely on the past history of messages received into that synaptic layer. They are known to change with temperature, and may well change with many other things.

This view of the nervous system corresponds to the theory of those machines that consist in a sequence of switching devices in which the opening of a later switch depends on the action of precise combinations of earlier switches leading into it, which open at the same time. This all-or-none machine is called a *digital* machine. It has great advantages for the most varied problems of communication and control. In particular, the sharpness of the decision between "yes" and "no" permits it to accumulate information in such a way as to allow us to discriminate very small differences in very large numbers.

Besides these machines which work on a yes-and-no scale, there are other computing and control machines which measure rather than count. These are known as *analogy* machines, because they operate on the basis of analogous connections between the measured quantities and the numerical quantities supposed to represent them. An example of an analogy machine is a slide rule, in contrast with a desk computing machine which operates digitally. Those who have used a slide rule know that the scale on which the marks have to be printed and the accuracy of our eyes give sharp limits to the precision with which the rule can be read. These limits are not as easily extended as one might think, by making the slide rule larger. A ten-foot slide rule will give only one decimal place more accuracy than a one-foot slide rule, and in order to do this, not only must each foot of the larger slide rule be constructed with the same precision as the smaller one, but the orientation of these successive feet must conform to the degree of accuracy to be expected for each one-foot slide rule. Furthermore, the problems of keeping the larger rule rigid are much greater than those which we find in the case of the smaller rule, and serve to limit the increase in accuracy which we get by increasing the size. In other words, for practical purposes, machines that measure, as opposed to machines that count, are very greatly limited in their precision. Add this to the prejudices of the physiologist in favor of all-or-none action, and we see why the greater part of the work which has been done on the mechanical simulacra of the brain has been on machines which are more or less on a digital basis.

However, if we insist too strongly on the brain as a glorified digital machine, we shall be subject to some very just criticism, coming in part from the physiologists and in part from the somewhat opposite camp of those psychologists who prefer not to make use of the machine comparison. I have said that in a digital machine there is a *taping*, which determines the sequence of operations to be performed, and that a change in this taping on the basis of past experience corresponds to a learning process. In the brain, the clearest analogy to taping is the determination of the synaptic thresholds, of the precise combinations of the incoming neurons which will fire an outgoing neuron with which they are connected. We have already seen that these thresholds are variable with temperature, and we have no reason to believe that they may not be variable with the chemistry of the blood and with many other phenomena which are not themselves originally of an all-or-none nature. It is therefore necessary that in considering the problem of learning, we should be most wary of assuming an all-or-none theory of the nervous system, without having made an intellectual criticism of the notion, and without specific experimental evidence to back our assumption.

It will often be said that there is no theory of learning whatever that will be reasonable for the machine. It will also be said that in the present stage of our knowledge, any theory of learning which I may offer will be premature, and will probably not correspond to the actual functioning of the nervous system. I wish to walk a middle path between these two criticisms. On the one hand, I wish to give a method of constructing learning machines, a method which will not only enable me to build certain special machines of this type, but will give me a general engineering technique for constructing a very large class of such machines. Only if I reach this degree of generality will I have defended myself in some measure from the criticism that the mechanical process which I claim is similar to learning, is, in fact, something of an essentially different nature from learning.

On the other hand, I wish to describe such machines in terms which are not too foreign to the actual observables of the nervous system, and of human and animal conduct. I am quite aware that I cannot expect to be right in detail in presenting the actual human mechanism, and that I may even be wrong in principle. Nevertheless, if I give a device which can be verbally formulated in terms of the concepts belonging to the human mind and the human brain, I shall give a point of departure for criticism, and a standard with which to compare the performance to be expected on the basis of other theories.

Locke, at the end of the seventeenth century, considered that the content of the mind was made up of what he calls *ideas*. The mind for him is entirely passive, a clean blackboard, *tabula rasa*, on which the experiences of the individual write their own impressions. If these impressions appear often, either under circumstances of simultaneity, or in a certain sequence, or in situations which we ordinarily attribute to cause and effect, then according to Locke, these impressions or ideas will form complex ideas, with a certain positive tendency for the component elements to stick together. The mechanism by which the ideas stick together lies in the ideas themselves; but there is throughout Locke's writing a singular unwillingness to describe such a mechanism. His theory can bear only the sort of relation to reality that a picture of a locomotive bears to a working locomotive. It is a diagram without any working parts. This is not remarkable when we consider the date of Locke's theory. It was in astronomy, and not in engineering or in psychology, that the dynamic point of view, the point of view of working parts, first reached its importance; and this was at the hands of Newton, who was not a predecessor of Locke, but a contemporary.

For several centuries, science, dominated by the Aristotelian impulse to classify, neglected the modern impulse to search for ways in which phenomena function. Indeed, with the plants and animals yet to be explored, it is hard to see how biological science could have entered a properly dynamic period except through the continual gathering of more descriptive natural history. The great botanist Linnaeus will serve us as an example. For Linnaeus, species and genera were fixed Aristotelian forms, rather than signposts for a process of evolution; but it was only on the basis of a thoroughly Linnaean description that any cogent case could ever be made for evolution. The early natural historians were the practical frontiersmen of the intellect; too much under the compulsion to seize and occupy new territory to be very precise in treating the problem of explaining the new forms that they had observed. After the frontiersman comes the operative farmer, and after the naturalist comes the modern scientist.

In the last quarter of the last century and the first quarter of the present one, another great scholar, Pavlov, covered in his own way essentially the same ground that Locke had covered earlier. His study of the conditioned reflexes, however, progressed experimentally, not theoretically as Locke's had. Moreover, he treated it as it appears among the lower animals rather than as it appears in man. The lower animals cannot speak in man's language, but in the language of behavior. Much of their more conspicuous behavior is emotional in its motivation and much of their emotion is concerned with food. It was with food that Pavlov began, and with the physical symptom of salivation. It is easy to insert a canula into the salivary duct of a dog and to observe the secretion that is stimulated by the presence of food.

Ordinarily many things unconnected with food, as objects seen, sounds heard, etc., produce no effect on salivation, but Pavlov observed that if a certain pattern or a certain sound had been systematically introduced to a dog at feeding time, then the display of the pattern or sound alone was sufficient to excite salivation. That is, the reflex of salivation was conditioned by a past association.

Here we have on the level of the animal reflex, something analogous to Locke's association of ideas, an association which occurs in reflex responses whose emotional content is presumably very strong. Let us notice the rather complicated nature of the antecedents which are needed to produce a conditioned reflex of the Pavlov type. To begin with, they generally center about something important to the life of the animal: in this case, food, even though in the reflex's final form the food element may be entirely elided. We may, however, illustrate the importance of the initial stimulus of a Pavlovian conditioned reflex by the example of electric fences enclosing a cattle farm.

On cattle farms, the construction of wire fences strong enough to turn a steer is not easy. It is thus economical to replace a heavy fence of this type by one where one or two relatively thin strands of wire carry a sufficiently high electric voltage to impress upon an animal a quite appreciable shock when the animal short-circuits it by contact with its body. Such a fence may have to resist the pressure of the steer once or twice; but after that, the fence acts, not because it can hold up mechanically under pressure, but because the steer has developed a conditioned reflex which tends to prevent it from coming into contact with the fence at all. Here the original trigger to the reflex is pain; and the withdrawal from pain is vital for the continued life of any animal. The transferred trigger is the sight of the fence. There are other triggers which lead to conditioned reflexes besides hunger and pain. It will be using anthropomorphic language to call these emotional situations, but there is no such anthropomorphism needed to describe them as situations which generally carry an emphasis and importance not belonging to many other animal experiences. Such experiences, whether we may call them emotional or not, produce strong reflexes. In the formation of conditioned reflexes in general the reflex response is transferred to one of these trigger situations. This trigger situation is one which frequently occurs concurrently with the original trigger. The change in the stimulus for which a given response takes place must have some such nervous correlate as the opening of a synaptic pathway leading to the response which would otherwise have been closed, or the closing of one which would otherwise have been open; and thus constitutes what Cybernetics calls a *change in taping*.

Such a change in taping is preceded by the continued association of the old, strong, natural stimulus for a particular reaction and the new concomitant one. It is as if the old stimulus had the power to change the permeability of those pathways which were carrying a message at the same time as it was active. The interesting thing is that the new, active stimulus need have almost nothing predetermined about it except the fact of repeated concomitance with the original stimulus. Thus the original stimulus seems to produce a long-time effect in all those pathways which were carrying a message at the time of its occurrence or at least in a large number of them. The insignificance of the substitute stimulus indicates that the modifying effect of the original stimulus is widespread, and is not confined to a few special pathways. Thus we assume that there may be some kind of general message released by the original stimulus, but that it is active only in those channels which were carrying a message at about the time of the original stimulus. The effect of this action may perhaps not be permanent, but is at least fairly long-lived. The most logical site at which to suppose this secondary action to take place is in the synapses, where it most probably affects their thresholds.

The concept of an undirected message spreading out until it finds a receiver, which is then stimulated by it, is not an unfamiliar one. Messages of this sort are used very frequently as alarms. The fire siren is a call to all the citizens of the town, and in particular to members of the fire department, wherever they may be. In a mine, when we wish to clear out all remote passages because of the presence of fire damp, we break a tube of ethyl mercaptan in the air-intake. There is no reason to suppose that such messages may not occur in the nervous system. If I were to construct a learning machine of a general type, I would be very much disposed to employ this method of the conjunction of general spreading "To-whom-it-may-concern" messages with localized channeled messages. It ought not to be too difficult to devise electrical methods of performing this task. This is very different, of course, from saying that learning in the animal actually occurs by such a conjunction of spreading and of channeled messages. Frankly, I think it is quite possible that it does, but our evidence is as yet not enough to make this more than a conjecture.

As to the nature of these "To-whom-it-may-concern" messages, supposing them to exist, I am on still more speculative ground. They might indeed be nervous, but I am rather inclined to attribute them to the non-digital, analogy side of the mechanism responsible for reflexes and thought. It is a truism to attribute synaptic action to chemical phenomena. Actually, in the action of a nerve, it is impossible to separate chemical potentials and electrical potentials, and the statement that a certain particular action is chemical is almost devoid of meaning. Nevertheless, it does no violence to current thought to suppose that at least one of the causes or concomitants of synaptic change is a chemical change which manifests itself locally, no matter what its origin may be. The presence of such a change may very well be locally dependent on release signals which are transmitted nervously. It is at least equally conceivable that changes of the sort may be due in part to chemical changes transmitted generally through the blood, and not by the nerves. It is conceivable that "To-whom-it-may-concern" messages are transmitted nervously, and make themselves locally apparent in the form of that sort of chemical action which accompanies synaptic changes. To me, as an engineer, the transmission of "To-whom-it-may-concern" messages would appear to be more economically performed through the blood than through the nerves. However, I have no evidence.

Let us remember that these "To-whom-it-may-concern" influences bear a certain similarity to the sort of changes in the anti-aircraft control apparatus which carry all new statistics to the instrument, rather than to those which directly carry only specific numerical data. In both cases, we have an action which has probably been piling up for a long time, and which will produce effects due to continue for a long time.

The rapidity with which the conditioned reflex responds to its stimulus is not necessarily an index that the conditioning of the reflex is a process of comparable speed. Thus it seems to me appropriate for a message causing such a conditioning to be carried by the slow but pervasive influence of the blood stream.

It is already a considerable narrowing of what my point of view requires, to suppose that the fixing influence of hunger or pain or whatever stimulus may determine a conditioned reflex passes through the blood. It would be a still greater restriction if I should try to specify the nature of this unknown blood-borne influence, if any such exists. That the blood carries in it substances which may alter nervous action directly or indirectly seems to me very likely, and to be suggested by the actions of some at least of the hormones or internal secretions. This, however, is not the same as saying that the influence on thresholds which determines learning is the product of specific hormones. Again, it is tempting to find the common denominator of hunger and the pain caused by the electrified fence in something that we may call an emotion, but it is certainly going too far to attach emotion to all conditioners of reflexes, without any further discussion of their particular nature.

Nevertheless, it is interesting to know that the sort of phenomenon which is recorded subjectively as emotion may not be merely a useless epiphenomenon of nervous action, but may control some essential stage in learning, and in other similar processes. I definitely do not say that it does, but I do say that those psychologists who draw sharp and uncrossable distinctions between man's emotions and those of other living organisms and the responses of the modern type of automatic mechanisms, should be just as careful in their denials as I should be in my assertions.

IV

THE MECHANISM AND HISTORY
OF LANGUAGE

Naturally, no theory of communication can avoid the discussion of language. Language, in fact, is in one sense another name for communication itself, as well as a word used to describe the codes through which communication takes place. We shall see later in this chapter that the use of encoded and decoded messages is important, not merely for human beings, but for other living organisms, and for the machines used by human beings. Birds communicate with one another, monkeys communicate with one another, insects communicate with one another, and in all this communication some use is made of signals or symbols which can be understood only by being privy to the system of codes involved.

What distinguishes human communication from the communication of most other animals is (a) the delicacy and complexity of the code used, and (b) the high degree of arbitrariness of this code. Many animals can signal their emotions to one another, and in signaling these emotions indicate the presence of an enemy, or of an animal of the same species but of opposite sex, and quite a variety of detailed messages of this sort. Most of these messages are fugitive and unstored. The greater part would be translated in human language into expletives and exclamations, although some might be rendered crudely by words to which we should be likely to give the form of nouns and adjectives, but which would be used by the animal in question without any corresponding distinction of grammatical form. In general, one would expect the language of animals to convey emotions first, things next, and the more complicated relations of things not at all.

Besides this limitation of the language of animals as it concerns the character of what is communicated, their language is very generally fixed by the species of the animal, and unchanging in history. One lion's roar is very nearly another lion's roar. Yet there are animals such as the parrot, the myna, and the crow, which seem to be able to pick up sounds from the surrounding environment, and particularly from the cries of other animals and of man, and to be able to modify or to augment their vocabularies, albeit within very narrow limits. Yet even these do not seem to have anything like man's freedom to use any pronounceable sound as a code for some meaning or other, and to pass on this code to the surrounding group in such a way that the codification forms an accepted language understood within the group, and almost meaningless on the outside.

Within their very great limitations, the birds that can imitate human speech have several characteristics in common: they are social, they are rather long-lived, and they have memories which are excellent by anything less than the exacting human standard. There is no doubt that a talking bird can learn to use human or animal sounds at the appropriate cues, and with what will appear at least to the casual listener as some element of understanding. Yet even the most vocal members of the sub-human world fail to compete with man in ease of giving significance to new sounds, in repertory of sounds carrying a specific codification in extent of linguistic memory, and above all in the ability to form symbols for relations, classes, and other entities, of Russell's "higher logical type."

I wish to point out nevertheless that language is not exclusively an attribute of living beings but one which they may share to a certain degree with the machines man has constructed. I wish to show further that man's preoccupation with language most certainly represents a possibility which is built into him, and which is not built into his nearest relatives, the great apes. Nevertheless, I shall show that it is built in only as a possibility which must be made good by learning.

We ordinarily think of communication and language as being directed from person to person. However, it is quite possible for a person to talk to a machine, a machine to a person, and a machine to a machine. For example, in the wilder stretches of our own West and of Northern Canada, there are many possible power sites far from any settlement where the workers can live, and too small to justify the foundation of new settlements on their own account, though not so small that the power systems are able to neglect them. It is thus desirable to operate these stations in a way that does not involve a resident staff, and in fact leaves the stations unattended for months between the rounds of a supervising engineer.

To accomplish this, two things are necessary. One of these is the introduction of automatic machinery; making it impossible to switch a generator on to a busbar or connecting member until it has come into the right frequency, voltage, and phase; and providing in a similar manner against other disastrous electrical, mechanical, and hydraulic contingencies. This type of operation would be enough if the daily cycle of the station were unbroken and unalterable.

This, however, is not the case. The load on a generating system depends on many variable factors. Among these are the fluctuating industrial demand; emergencies which may remove a part of the system from operation; and even passing clouds, which may make tens of thousands of offices and homes turn on their electric lights in the middle of the day. It follows that the automatic stations, as well as those operated by a working crew, must be within constant reach of the load dispatcher, who must be able to give orders to his machines; and this he does by sending appropriately coded signals to the power station, either over a special line designed for the purpose, or over existing telegraph or telephone lines, or over a carrier system making use of the power lines themselves. On the other hand, before the load dispatcher can give his orders intelligently, he must be acquainted with the state of affairs at the generating station. In particular, he must know whether the orders he has given have been executed, or have been held up through some failure in the equipment. Thus the machines in the generating station must be able to send return messages to the load dispatcher. Here, then, is one instance of language emanating from man and directed toward the machine, and vice versa.

It may seem curious to the reader that we admit machines to the field of language and yet almost totally deny language to the ants. Nevertheless, in constructing machines, it is often very important for us to extend to them certain human attributes which are not found among the lower members of the animal community. If the reader wishes to conceive this as a metaphoric extension of our human personalities, he is welcome to do so; but he should be cautioned that the new machines will not stop working as soon as we have stopped giving them human support.

The language directed toward the machine actually consists of more than a single step. From the point of view of the line engineer alone, the code transmitted along the line is complete in itself. To this message we may apply all the notions of Cybernetics, or the theory of messages. We may evaluate the amount of information it carries by determining its probability in the ensemble of all possible messages, and then taking the negative logarithm of this probability, in accordance with the theory expounded in Chapter I. However, this represents not the information actually carried by the line, but the maximum amount it might carry, if it were to lead into proper terminal equipment. The amount of information carried with actual terminal equipment depends on the ability of the latter to transmit or to employ the information received.

We are thus led to a new conception of the way in which the generating station receives the orders. Its actual performance of opening and closing switches, of pulling generators into phase, of controlling the flow of water in sluices, and of turning the turbines on or off, may be regarded as a language in itself, with a system of probabilities of behavior given by its own history. Within this frame every possible sequence of orders has its own probability, and hence carries its own amount of information.

It is, of course, possible that the relation between the line and the terminal machine is so perfect that the amount of information contained in a message, from the point of view of the carrying capacity of the line, and the amount of information of the fulfilled orders, measured from the point of view of the operation of the machine, will be identical with the amount of information transmitted over the compound system consisting of the line followed by the machine. In general, however, there will be a stage of translation between the line and the machine; and in this stage, information may be lost which can never be regained. Indeed, the process of transmitting information may involve several consecutive stages of transmission following one another in addition to the final or effective stage; and between any two of these there will be an act of translation, capable of dissipating information. That information may be dissipated but not gained, is, as we have seen, the cybernetic form of the second law of thermodynamics.

Up to this point in this chapter we have been discussing communication systems terminating in machines. In a certain sense, all communication systems terminate in machines, but the ordinary language systems terminate in the special sort of machine known as a human being. The human being as a terminal machine has a communication network which may be considered at three distinct levels. For ordinary spoken language, the first human level consists of the ear, and of that part of the cerebral mechanism which is in permanent and rigid connection with the inner ear. This apparatus, when joined to the apparatus of sound vibrations in the air, or their equivalent in electric circuits, represents the machine concerned with the *phonetic* aspect of language, with sound itself.

The *semantic* or second aspect of language is concerned with meaning, and is apparent, for example, in difficulties of translating from one language to another where the imperfect correspondence between the meanings of words restricts the flow of information from one into the other. One may get a remarkable semblance of a language like English by taking a sequence of words, or pairs of words, or triads of words, according to the statistical frequency with which they occur in the language, and the gibberish thus obtained will have a remarkably persuasive similarity to good English. This meaningless simulacrum of intelligent speech is practically equivalent to significant language from the phonetic point of view, although it is semantically balderdash, while the English of an intelligent foreigner whose pronunciation is marked by the country of his birth, or who speaks literary English, will be semantically good and phonetically bad. On the other hand, the average synthetic after-dinner speech is phonetically good and semantically bad.

In human communication apparatus, it is possible but difficult to determine the characteristics of its phonetic mechanism, and therefore also possible but difficult to determine what is phonetically significant information, and to measure it. It is clear, for example, that the ear and the brain have an effective frequency cutoff preventing the reception of some high frequencies which can penetrate the ear and can be transmitted by the telephone. In other words, these high frequencies, whatever information they may give an appropriate receptor, do not carry any significant amount of information for the ear. But it is even more difficult to determine and measure semantically significant information.

Semantic reception demands memory, and its consequent long delays. The types of abstractions belonging to the important semantic stage are not merely those associated with built-in permanent subassemblies of neurons in the brain, such as those which must play a large role in the perception of geometrical form; but with abstraction-detector-apparatus consisting of parts of the *internuncial pool*—that is, of sets of neurons which are available for larger assemblies, but are not permanently locked into them—which have been temporarily assembled for the purpose.

Besides the highly organized and permanent assemblies in the brain that undoubtedly exist, and are found in those parts of the brain associated with the organs of special sense, as well as in other places, there are particular switchings and connections which seem to have been formed temporarily for special purposes, such as learned reflexes and the like. In order to form such particular switchings, it must be possible to assemble sequences of neurons available for the purpose that are not already in use. This question of assembling concerns, of course, the synaptic thresholds of the sequence of neurons assembled. Since neurons exist which can either be within or outside of such temporary assemblies, it is desirable to have a special name for them. As I have already indicated, I consider that they correspond rather closely to what the neurophysiologists call internuncial pools.

This is at least a reasonable theory of their behavior.

The semantic receiving apparatus neither receives nor translates the language word by word, but idea by idea, and often still more generally. In a certain sense, it is in a position to call on the whole of past experience in its transformations, and these long-time carry-overs are not a trivial part of its work.

There is a third level of communication, which represents a translation partly from the semantic level and partly from the earlier phonetic level. This is the translation of the experiences of the individual, whether conscious or unconscious, into actions which may be observed externally. We may call this *the behavior level* of language. In the lower animals, it is the only level of language which we may observe beyond the phonetic input. Actually this is true in the case of every human being other than the particular person to whom any given passage is addressed in each particular case; in the sense that that person can have access to the internal thoughts of another person only through the actions of the latter. These actions consist of two parts: namely, direct gross actions, of the sort which we also observe in the lower animals; and in the coded and symbolic system of actions which we know of as spoken or written language.

It is theoretically not impossible to develop the statistics of the semantic and behavior languages to such a level that we may get a fair measure of the amount of information that they contain. Indeed we can show by general observations that phonetic language reaches the receiving apparatus with less overall information than was originally sent, or at any rate with not more than the transmission system leading to the ear can convey; and that both semantic and behavior language contain less information still. This fact again is a corollary of the second law of thermodynamics, and is necessarily true if at each stage we regard the information transmitted as the maximum information that could be transmitted with an appropriately coded receiving system.

Let me now call the attention of the reader to something which he may not consider a problem at all— namely, the reason that chimpanzees do not talk. The behavior of chimpanzees has for a long time been a puzzle to those psychologists who have concerned themselves with these interesting beasts. The young chimpanzee is extraordinarily like a child, and clearly his equal or perhaps even his superior in intellectual matters. The animal psychologists have not been able to keep from wondering why a chimpanzee brought up in a human family and subject to the impact of human speech until the age of one or two, does not accept language as a mode of expression, and itself burst into baby talk.

Fortunately, or unfortunately as the case may be, most chimpanzees, in fact all that have as yet been observed, persist in being good chimpanzees, and do not become quasi-human morons. Nevertheless I think that the average animal psychologist is rather longingly hoping for that chimpanzee who will disgrace his simian ancestry by adhering to more human modes of conduct. The failure so far is not a matter of sheer bulk of intelligence, for there are defective human animals whose brains would shame a chimpanzee. It just does not belong to the nature of the beast to speak, or to want to speak.

Speech is such a peculiarly human activity that it is not even approached by man's closest relatives and his most active imitators. The few sounds emitted by chimpanzees have, it is true, a great deal of emotional content, but they have not the fineness of clear and repeated accuracy of organization needed to make them into a code much more accurate than the yowlings of a cat. Moreover (and this differentiates them still more from human speech), at times they belong to the chimpanzee as an unlearned inborn manifestation, rather than as the learned behavior of a member of a given social community.

The fact that speech belongs in general to man as man, but that a particular form of speech belongs to man as a member of a particular social community, is most remarkable. In the first place, taking the whole wide range of man as we know him today, it is safe to say that there is no community of individuals, not mutilated by an auditory or a mental defect, which does not have its own mode of speech. In the second place, all modes of speech are learned, and notwithstanding the attempts of the nineteenth century to formulate a genetic evolutionistic theory of languages, there is not the slightest general reason to postulate any single native form of speech from which all the present forms are originated. It is quite clear that if left alone, babies will make attempts at speech. These attempts, however, show their own inclinations to utter something, and do not follow any existing form of language. It is almost equally clear that if a community of children were left out of contact with the language of their seniors through the critical speech-forming years, they would emerge with something, which crude as it might be, would be unmistakably a language.

Why is it then that chimpanzees cannot be forced to talk, and that human children cannot be forced not to? Why is it that the general tendencies to speak and the general visual and psychological aspects of language are so uniform over large groups of people, while the particular linguistic manifestation of these aspects is varied? At least partial understanding of these matters is essential to any comprehension of the language-based community. We merely state the fundamental facts by saying that in man, unlike the apes, the impulse to use some sort of language is overwhelming; but that the particular language used is a matter which has to be learned in each special case. It apparently is built into the brain itself, that we are to have a preoccupation with codes and with the sounds of speech, and that the preoccupation with codes can be extended from those dealing with speech to those that concern themselves with visual stimuli. However, there is not one fragment of these codes which is born into us as a pre-established ritual, like the courting dances of many of the birds, or the system by which ants recognize and exclude intruders into the nest. The gift of speech does not go back to a universal Adamite language disrupted in the Tower of Babel. It is strictly a psychological impulse, and is not the gift of speech, but the gift of the power of speech.

In other words, the block preventing young chimpanzees from learning to talk is a block which concerns the semantic and not the phonetic stage of language. *The chimpanzee has simply no built-in mechanism which leads it to translate the sounds that it hears into the basis around which to unite its own ideas or into a complex mode of behavior.* Of the first of these statements we cannot be sure because we have no direct way of observing it. The second is simply a noticeable empirical fact. It may have its limitations, but that there is such a built-in mechanism in man is perfectly clear.

In this book, we have already emphasized man's extraordinary ability to learn as a distinguishing characteristic of the species, which makes social life a phenomenon of an entirely different nature from the apparently analogous social life among the bees and ants and other social insects. The evidence concerning children who have been deprived of contact with their own race over the years normally critical in the ordinary acquisition of language, is perhaps not completely unambiguous. The "Wolf Child" stories, which have led to Kipling's imaginative *Jungle Books*, with their public-school bears and Sandhurst wolves, are almost as little to be relied on in their original stark squalidity as in the *Jungle Books* idealizations. However, what evidence there is goes to show that there is a critical period during which speech is most readily learned; and that if this period is passed over without contact with one's fellow human beings, of whatever sort they may be, the learning of language becomes limited, slow, and highly imperfect.

This is probably true of most other abilities which we consider natural skills. If a child does not walk until it is three or four years old, it may have lost all the desire to walk. Ordinary locomotion may become a harder task than driving a car for the normal adult. If a person has been blind from childhood, and the blindness has been resolved by a cataract operation or the implantation of a transparent corneal section, the vision that ensues will, for a time, certainly bring nothing but confusion to those activities which have normally been carried out in darkness. This vision may never be more than a carefully learned new attainment of doubtful value. Now, we may fairly take it that the whole of human social life in its normal manifestations centers about speech, and that if speech is not learned at the proper time, the whole social aspect of the individual will be aborted.

To sum up, the human interest in language seems to be an innate interest in coding and decoding, and this seems to be as nearly specifically human as any interest can be. *Speech is the greatest interest and most distinctive achievement of man.*

I was brought up as the son of a philologist, and questions concerning the nature and technique of language have interested me from my childhood. It is impossible for as thorough a revolution in the theory of language as is offered by modern communication theory to take effect without effecting past linguistic ideas. As my father was a very heretical philologist whose influence tended to lead philology in much the same direction as the modern influences of communication theory, I wish to continue this chapter with a few amateurish reflections on the history of language and the history of our theory of language.

Man has held the notion that language is a mystery since very early times. The riddle of the Sphinx is a primitive conception of wisdom. Indeed, the very word riddle is derived from the root "to rede," or to puzzle out. Among many primitive people writing and sorcery are not far apart. The respect for writing goes so far in some parts of China that people are loath to throw away scraps of old newspapers and useless fragments of books.

Close to all these manifestations is the phenomenon of "name magic" in which members of certain cultures go from birth to death under names that are not properly their own, in order that they may not give a sorcerer the advantage of knowing their true names. Most familiar to us of these cases is that of the name of Jehovah of the Jews, in which the vowels are taken over from that other name of God, "Adonai," so that the Name of Power may not be blasphemed by being pronounced in profane mouths.

From the magic of names it is but a step to a deeper and more scientific interest in language. As an interest in textual criticism in the authenticity of oral traditions and of written texts it goes back to the ancients of all civilizations. A holy text must be kept pure. When there are divergent readings they must be resolved by some critical commentator. Accordingly, the Bible of the Christians and the Jews, the sacred books of the Persians and the Hindus, the Buddhist scriptures, the writings of Confucius, all have their early commentators. What has been learned for the maintenance of true religion has been carried out as a literary discipline, and textual criticism is one of the oldest of intellectual studies.

For a large part of the last century philological history was reduced to a series of dogmas which at times show a surprising ignorance of the nature of language.

The model of the Darwinian evolutionism of the times was taken too seriously and too uncritically. As this whole subject depends in the most intimate manner on our views of the nature of communication, I shall comment on it at a certain length.

The early speculation that Hebrew was the language of man in Paradise, and that the confusion of language originated at the building of the Tower of Babel, need not interest us here as anything more than a primitive precursor of scientific thought. However, the later developments of philological thought have retained for a long time a similar naïveté. That languages are related, and that they undergo progressive changes leading in the end to totally different languages, were observations which could not long remain unnoticed by the keen philological minds of the Renaissance. A book such as Ducange's *Glossarium Mediae atque Infimae Latinitatis* could not exist without its being clear that the roots of the Romance languages are not only in Latin, but in vulgar Latin. There must have been many learned rabbis who were well aware of the resemblance of Hebrew, Arabic, and Syriac. When, under the advice of the much maligned Warren Hastings, the East India Company founded its School of Oriental Studies at Fort William, it was no longer possible to ignore that Greek and Latin on the one hand, and Sanskrit on the other, were cut from the same cloth. At the beginning of the last century the work of the brothers Grimm and of the Dane, Rask, showed not only that the Teutonic languages came within the orbit of this so-called Indo-European group, but went further to make clear the linguistic relations of these languages to one another, and to a supposed distant common parent.

Thus evolutionism in language antedates the refined Darwinian evolutionism in biology. Valid as this evolutionism is, it very soon began to outdo biological evolutionism in places where the latter was not applicable. It assumed, that is, that the languages were independent, quasi-biological entities, with their developments modified entirely by internal forces and needs. In fact, they are epiphenomena of human intercourse, subject to all the social forces due to changes in the pattern of that intercourse.

In the face of the existence of *Mischsprachen*, of languages such as Lingua Franca, Swahili, Yiddish, Chinook Jargon, and even to a considerable extent English, there has been an attempt to trace each language to a single legitimate ancestor, and to treat the other participants in its origin as nothing more than godparents of the newborn child. There has been a scholars' distinction between legitimate phonetic formations following accepted laws, and such regrettable accidents as nonce words, popular etymologies, and slang. On the grammatical side, the original attempt to force all languages of any origin whatsoever into the strait jacket manufactured for Latin and Greek has been succeeded by an attempt almost as rigorous to form for each of them its own paradigms of construction.

It is scarcely until the recent work of Otto Jespersen that any considerable group of philologists have had objectivity enough to make of their science a representation of language as it is actually spoken and written, rather than a copybook attempt to teach the Eskimos how to speak Eskimo, and the Chinese how to write Chinese. The effects of misplaced grammatical purism are to be seen well outside of the schools. First among these, perhaps, is the way in which the Latin language, like the earlier generation of classical gods, has been slain by its own children.

During the Middle Ages Latin of a varying quality, the best of it quite acceptable to anyone but a pedant, remained the universal language of the clergy and of all learned men throughout Western Europe, even as Arabic has remained in the Moslem world down to the present day. This continued prestige of Latin was made possible by the willingness of writers and speakers of the language either to borrow from other languages, or to construct within the frame of Latin itself, all that was necessary for the discussion of the live philosophical problems of the age. The Latin of Saint Thomas is not the Latin of Cicero, but Cicero would have been unable to discuss Thomistic ideas in the Ciceronian Latin.

It may be thought that the rise of the vulgar languages of Europe must necessarily have marked the end of the function of Latin. This is not so. In India, notwithstanding the growth of the neo-Sanskritic languages, Sanskrit has shown a remarkable vitality lasting down to the present day. The Moslem world, as I have said, is united by a tradition of classical Arabic, even though the majority of Moslems are not Arabic speakers and the spoken Arabic of the present day has divided itself into a number of very different dialects. It is quite possible for a language which is no longer the language of vulgar communication to remain the language of scholarship for generations and even for centuries. Modern Hebrew has survived for two thousand years the lack of use of Hebrew in the time of Christ, and indeed has come back as a modern language of daily life. In what I am discussing now, I am referring only to the limited use of Latin as a language of learned men.

With the coming of the Renaissance, the artistic standards of the Latinists became higher, and there was more and more a tendency to throw out all post-classical neologisms. In the hands of the great Italian scholars of the Renaissance, this reformed Latin could be, and often was, a work of art; but the training necessary to wield such a delicate and refined tool was beyond that which would be incidental to the training of the scientist, whose main work must always concern itself with content rather than with perfection of form.

The result was that the people who *taught* Latin and the people who *used* Latin became ever more widely separated classes, until the teachers completely eschewed the problem of teaching their disciples anything but the most polished and unusable Ciceronian speech. In this vacuum they ultimately eliminated any function for themselves other than that of specialists; and as the specialty of Latinism thus came to be less and less in general demand, they abolished their own function. For this sin of pride, we now have to pay in the absence of an adequate international language far superior to the artificial ones such as Esperanto, and well suited for the demands of the present day.

Alas, the attitudes of the classicists are often beyond the understanding of the intelligent layman! I recently had the privilege of hearing a commencement address from a classicist who bewailed the increased centrifugal force of modern learning, which drives the natural scientist, the social scientist, and the literary man ever farther from one another. He put it into the form of an imaginary trip which he took through a modern university, as the guide and mentor to a reincarnated Aristotle. His talk began by presenting in the pillory bits of technical jargon from each modern intellectual field, which he supposed himself to have presented to Aristotle as horrible examples. May I remark that all we possess of Aristotle is what amounts to the school notebooks of his disciples, written in one of the most crabbed technical jargons in the history of the world, and totally unintelligible to any contemporary Greek who had not been through the discipline of the Lyceum? That this jargon has been sanctified by history, so that it has become itself an object of classical education, is not relevant; for this happened after Aristotle, not contemporaneously with him. The important thing is that the Greek language of the time of Aristotle was ready to compromise with the technical jargon of a brilliant scholar, while even the English of his learned and reverend successors is not willing to compromise with the similar needs of modern speech.

With these admonitory words, let us return to a modern point of view which assimilates the operation of linguistic translation and the related operations of the interpretation of language by ear and by brain to the performance and the coupling of non-human communication networks. It will be seen that this is really in accordance with the modern and once heretical views of Jespersen and his school. Grammar is no longer primarily normative. It has become factual. The question is not what code should we use, but what code do we use. It is quite true that in the finer study of language, normative questions do indeed come into play, and are very delicate. Nevertheless, they represent the last fine flower of the communication problem, and not its most fundamental stages.

We have thus established the basis in man for the simplest element of his communication: namely, the communication of man with man by the immediate use of language, when two men are face to face with one another. The inventions of the telephone, the telegraph, and other similar means of communication have shown that this capacity is not intrinsically restricted to the immediate presence of the individual, for we have many means to carry this tool of communication to the ends of the earth.

Among primitive groups the size of the community for an effective communal life is restricted by the difficulty of transmitting language. For many millennia, this difficulty was enough to reduce the optimum size of the state to something of the order of a few million people, and generally fewer. It will be noted that the great empires which transcended this limited size were held together by improved means of communication. The heart of the Persian Empire was the Royal Road and the relay of messengers who conveyed the Royal Word along it. The great empire of Rome was possible only because of Rome's progress in roadbuilding. These roads served to carry not only the legions, but the written authority of the Emperor as well. With the airplane and the radio the word of the rulers extends to the ends of the earth, and very many of the factors which previously precluded a World State have been abrogated. It is even possible to maintain that modern communication, which forces us to adjudicate the international claims of different broadcasting systems and different airplane nets, has made the World State inevitable.

But as efficient as communications' mechanisms become, they are still, as they have always been, subject to the overwhelming tendency for entropy to increase, for information to leak in transit, unless certain external agents are introduced to control it. I have already referred to an interesting view of language made by a cybernetically-minded philologist—that speech is a joint game by the talker and the listener against the forces of confusion. On the basis of this description, Dr. Benoit Mandelbrot has made certain computations concerning the distribution of the lengths of words in an optimal language, and has compared these results with what he has found in existing languages. Mandelbrot's results indicate that a language optimal according to certain postulates will very definitely exhibit certain distribution of length among words. This distribution is very different from what will be found in an artificial language, such as Esperanto or Volapük. On the other hand, it is remarkably close to what is found in most actual languages that have withstood the attrition of use for centuries. The results of Mandelbrot do not, it is true, give an absolutely fixed distribution of word lengths; in his formulas there still occur certain quantities which must be assigned, or, as the mathematician calls them, *parameters*. However, by a proper choice of these parameters, Mandelbrot's theoretical results fit very closely the word distribution in many actual languages, indicating that there is a certain natural selection among them, and that the form of a language which survives by the very fact of its use and survival has been driven to take something not too remotely resembling an optimum form of distribution.

The attrition of language may be due to several causes. Language may strive simply against nature's tendency to confuse it or against willful human attempts to subvert its meaning.[1] Normal communicative discourse, whose major opponent is the entropic tendency of nature itself, is not confronted by an active enemy, conscious of its own purposes. Forensic discourse, on the other hand, such as we find in the law court in legislative debates and so on, encounters a much more formidable opposition, whose conscious aim is to qualify and even to destroy its meaning. Thus an adequate theory of language as a game should distinguish between these two varieties of language, one of which is intended primarily to convey information and the other primarily to impose a point of view against a willful opposition. I do not know if any philologist has yet made the technical observations and theoretical propositions which are necessary to distinguish these two classes of language for our purposes, but I am quite sure that they are substantially different forms. I shall talk further about forensic language in a later chapter, which deals with language and law.

The desire to apply Cybernetics of semantics, as a discipline to control the loss of meaning from language, has already resulted in certain problems. It seems necessary to make some sort of distinction between information taken brutally and bluntly, and that sort of information on which we as human beings can act effectively or, *mutatis mutandis,* on which the machine can act effectively. In my opinion, the central distinction and difficulty here arises from the fact that it

[1] Relevant here also is Einstein's aphorism, see Chapter II, p. 35 above.

is not the quantity of information sent that is important for action, but rather the quantity of information which can penetrate into a communication and storage apparatus sufficiently to serve as the trigger for action.

I have said that any transmission of, or tampering with, messages decreases the amount of information they contain, unless new information is fed in, either from new sensations or from memories which have been previously excluded from the information system. This statement, we have seen, is another version of the second law of thermodynamics. Now let us consider an information system used to control the sort of electric power sub-station of which we spoke earlier in the chapter. What is important is not merely the information that we put into the line, but what is left of it when it goes through the final machinery to open or close sluices, to synchronize generators, and to do similar tasks. In one sense, this terminal apparatus may be regarded as a filter superimposed on the transmission line. Semantically significant information from the cybernetic point of view is that which gets through the line-plus-filter, rather than that which gets through the line alone. In other words, when I hear a passage of music, the greater part of the sound gets to my sense organs and reaches my brain. However, if I lack the perception and training necessary for the aesthetic understanding of musical structure, this information will meet a block, whereas if I were a trained musician it would meet an interpreting structure or organization which would exhibit the pattern in a significant form which can lead to aesthetic appreciation and further understanding. Semantically significant information in the machine as well as in man is information which gets through to an activating mechanism in the system that receives it, despite man's and/or nature's attempts to subvert it. From the point of view of Cybernetics, semantics defines the extent of meaning and controls its loss in a communications system.

V

ORGANIZATION AS THE MESSAGE

The present chapter will contain an element of phantasy. Phantasy has always been at the service of philosophy, and Plato was not ashamed to clothe his epistemology in the metaphor of the cave. Dr. J. Bronowski among others has pointed out that mathematics, which most of us see as the most factual of all sciences, constitutes the most colossal metaphor imaginable, and must be judged, aesthetically as well as intellectually, in terms of the success of this metaphor.

The metaphor to which I devote this chapter is one in which the organism is seen as message. Organism is opposed to chaos, to disintegration, to death, as message is to noise. To describe an organism, we do not try to specify each molecule in it, and catalogue it bit by bit, but rather to answer certain questions about it which reveal its pattern: a pattern which is more significant and less probable as the organism becomes, so to speak, more fully an organism.

We have already seen that certain organisms, such as man, tend for a time to maintain and often even to increase the level of their organization, as a local enclave in the general stream of increasing entropy, of increasing chaos and de-differentiation. Life is an island here and now in a dying world. The process by which we living beings resist the general stream of corruption and decay is known as *homeostasis*.

We can continue to live in the very special environment which we carry forward with us only until we begin to decay more quickly than we can reconstitute ourselves. Then we die. If our bodily temperature rises or sinks one degree from its normal level of 98.6° Fahrenheit, we begin to take notice of it, and if it rises or sinks ten degrees, we are all but sure to die. The oxygen and carbon dioxide and salt in our blood, the hormones flowing from our ductless glands, are all regulated by mechanisms which tend to resist any untoward changes in their levels. These mechanisms constitute what is known as homeostasis, and are negative feedback mechanisms of a type that we may find exemplified in mechanical automata.

It is the pattern maintained by this homeostasis, which is the touchstone of our personal identity. Our tissues change as we live: the food we eat and the air we breathe become flesh of our flesh and bone of our bone, and the momentary elements of our flesh and bone pass out of our body every day with our excreta. We are but whirlpools in a river of ever-flowing water. We are not stuff that abides, but patterns that perpetuate themselves.

A pattern is a message, and may be transmitted as a message. How else do we employ our radio than to transmit patterns of sound, and our television set than to transmit patterns of light? It is amusing as well as instructive to consider what would happen if we were to transmit the whole pattern of the human body, of the human brain with its memories and cross connections, so that a hypothetical receiving instrument could re-embody these messages in appropriate matter, capable of continuing the processes already in the body and the mind, and of maintaining the integrity needed for this continuation by a process of homeostasis.

Let us invade the realm of science fiction. Some forty-five years ago, Kipling wrote a most remarkable little story. This was at the time when the flights of the Wright brothers had become familiar to the world, but before aviation was an everyday matter. He called this story "With the Night Mail," and it purports to be an account of a world like that of today, when aviation should have become a matter of course and the Atlantic a lake to be crossed in one night. He supposed that aerial travel had so united the world that war was obsolete, and that all the world's really important affairs were in the hands of an Aerial Board of Control, whose primary responsibility extended to air traffic, while its secondary responsibility extended to "all that that implies." In this way, he imagined that the various local authorities had gradually been compelled to drop their rights, or had allowed their local rights to lapse; and that the central authority of the Aerial Board of Control had taken these responsibilities over. It is rather a fascist picture which Kipling gives us, and this is understandable in view of his intellectual presuppositions, even though fascism is not a necessary condition of the situation which he envisages. His millennium is the millennium of a British colonel back from India. Moreover, with his love for the gadget as a collection of wheels that rotate and make a noise, he has emphasized the extended physical transportation of man, rather than the transportation of language and ideas. He does not seem to realize that where a man's word goes, and where his power of perception goes, to that point his control and in a sense his physical existence is extended. To see and to give commands to the whole world is almost the same as being everywhere. Given his limitations Kipling, nevertheless, had a poet's insight, and the situation he foresaw seems rapidly coming to pass.

To see the greater importance of the transportation of information as compared with mere physical transportation, let us suppose that we have an architect in Europe supervising the construction of a building in the United States. I am assuming, of course, an adequate working staff of constructors, clerks of the works, etc., on the site of the construction. Under these conditions, even without transmitting or receiving any material commodities, the architect may take an active part in the construction of the building. Let him draw up his plans and specifications as usual. Even at present, there is no reason why the working copies of these plans and specifications must be transmitted to the construction site on the same paper on which they have been drawn up in the architect's drafting-room. Ultrafax gives a means by which a facsimile of all the documents concerned may be transmitted in a fraction of a second, and the received copies are quite as good working plans as the originals. The architect may be kept up to date with the progress of the work by photographic records taken every day or several times a day; and these may be forwarded back to him by Ultrafax. Any remarks or advice he cares to give his representative on the job may be transmitted by telephone, Ultrafax, or teletypewriter. In short, the bodily transmission of the architect and his documents may be replaced very effectively by the message-transmission of communications which do not entail the moving of a particle of matter from one end of the line to the other.

If we consider the two types of communication: namely, material transport, and transport of information alone, it is at present possible for a person to go from one place to another only by the former, and not as a message. However, even now the transportation of messages serves to forward an extension of man's senses and his capabilities of action from one end of the world to another. We have already suggested in this chapter that the distinction between material transportation and message transportation is not in any theoretical sense permanent and unbridgeable.

This takes us very deeply into the question of human individuality. The problem of the nature of human individuality and of the barrier which separates one personality from another is as old as history. The Christian religion and its Mediterranean antecedents have embodied it in the notion of *soul*. The individual possesses a soul, so say the Christians, which has come into being by the act of conception, but which will continue in existence for all eternity, either among the Blessed or among the Damned, or in one of the little intermediate lacunae of Limbo which the Christian faith allows.

The Buddhists follow a tradition which agrees with the Christian tradition in giving to the soul a continuity after death, but this continuity is in the body of another animal or another human being, rather than in some Heaven or Hell. There are indeed Buddhist Heavens and Hells, although the stay of the individual there is generally temporary. In the most final Heaven of the Buddhists, however, the state of Nirvana, the soul loses its separate identity and is absorbed into the Great Soul of the World.

These views have been without the benefit of the influence of science. The most interesting early scientific account of the continuity of the soul is Leibnitz's which conceives the soul as belonging to a larger class of permanent spiritual substances which he called *monads*. These monads spend their whole existence from the creation on in the act of perceiving one another; although some perceive with great clarity and distinctness, and others in a blurred and confused manner. This perception does not however represent any true interaction of the monads. The monads "have no windows," and have been wound up by God at the creation of the world so that they shall maintain their foreordained relationships with one another through all eternity. They are indestructible.

Behind Leibnitz's philosophical views of the monads there lie some very interesting biological speculations. It was in Leibnitz's time that Leeuwenhoek first applied the simple microscope to the study of very minute animals and plants. Among the animals that he saw were spermatozoa. In the mammal, spermatozoa are infinitely easier to find and to see than ova. The human ova are emitted one at a time, and unfertilized uterine ova or very early embryos were until recently rarities in the anatomical collections. Thus the early miscroscopists were under the very natural temptation to regard the spermatozoon as the only important element in the development of the young, and to ignore entirely the possibility of the as yet unobserved phenomenon of fertilization. Furthermore, their imagination displayed to them in the front segment or head of the spermatozoon a minute fetus, rolled up with head forward. This fetus was supposed to contain in itself spermatozoa which were to develop into the next generation of fetuses and adults, and so on *ad infinitum.* The female was supposed to be merely the nurse of the spermatozoon.

Of course, from the modern point of view, this biology is simply false. The spermatozoon and the ovum are nearly equal participants in determining individual heredity. Furthermore, the germ cells of the future generation are contained in them *in posse,* and not *in esse.* Matter is not infinitely divisible, nor indeed from any absolute standpoint is it very finely divisible; and the successive diminutions required to form the Leeuwenhoek spermatozoon of a moderately high order would very quickly lead us down beyond electronic levels.

In the view now prevalent, as opposed to the Leibnitzian view, the continuity of an individual has a very definite beginning in time, but it may even have a termination in time quite apart from the death of the individual. It is well known that the first cell division of the fertilized ovum of a frog leads to two cells, which can be separated under appropriate conditions. If they are so separated, each will grow into a complete frog. This is nothing but the normal phenomenon of identical twinning in a case in which the anatomical accessibility of the embryo is sufficient to permit experimentation. It is exactly what occurs in human identical twins, and is the normal phenomenon in those armadillos that bear a set of identical quadruplets at each birth. It is the phenomenon, moreover, which gives rise to double monsters, when the separation of the two parts of the embryo is incomplete.

This problem of twinning may not however appear as important at first sight as it really is, because it does not concern animals or human beings with what may be considered well-developed minds and souls. Not even the problem of the double monster, the imperfectly separated twins, is too serious in this respect. Viable double monsters must always have either a single central nervous system or a well-developed pair of separate brains. The difficulty arises at another level in the problem of split personalities.

A generation ago, Dr. Morton Prince of Harvard gave the case history of a girl, within whose body several better-or-worse-developed personalities seemed to succeed one another, and even to a certain extent to coexist. It is the fashion nowadays for the psychiatrists to look down their noses a little bit when Dr. Prince's work is mentioned, and to attribute the phenomenon to hysteria. It is quite possible that the separation of the personalities was never as complete as Prince sometimes appears to have thought it to be, but for all that it was a separation. The word "hysteria" refers to a phenomenon well observed by the doctors, but so little explained that it may be considered but another question-begging epithet.

One thing at any rate is clear. The physical identity of an individual does not consist in the matter of which it is made. Modern methods of tagging the elements participating in metabolism have shown a much higher turnover than was long thought possible, not only of the body as a whole, but of each and every component part of it. The biological individuality of an organism seems to lie in a certain continuity of process, and in the memory by the organism of the effects of its past development. This appears to hold also of its mental development. In terms of the computing machine, the individuality of a mind lies in the retention of its earlier tapings and memories, and in its continued development along lines already laid out.

Under these conditions, just as a computing machine may be used as a pattern on which to tape other computing machines, and just as the future development of these two machines will continue parallel except for future changes in taping and experience, so too, there is no inconsistency in a living individual forking or divaricating into two individuals sharing the same past, but growing more and more different. This is what happens with identical twins; but there is no reason why it could not happen with what we call the mind, without a similar split of the body. To use computing-machine language again, at some stage a machine which was previously assembled in an all-over manner may find its connections divided into partial assemblies with a higher or lower degree of independence. This would be a conceivable explanation of Prince's observations.

Moreover, it is thinkable that two large machines which had previously not been coupled may become coupled so as to work from that stage on as a single machine. Indeed this sort of thing occurs in the union of the germ cells, although perhaps not on what we would ordinarily call a purely mental level. The mental identity necessary for the Church's view of the individuality of the soul certainly does not exist in any absolute sense which would be acceptable to the Church.

To recapitulate: the individuality of the body is that of a flame rather than that of a stone, of a form rather than of a bit of substance. This form can be transmitted or modified and duplicated, although at present we know only how to duplicate it over a short distance.

When one cell divides into two, or when one of the genes which carries our corporeal and mental birthright is split in order to make ready for a reduction division of a germ cell, we have a separation in matter which is conditioned by the power of a pattern of living tissue to duplicate itself. Since this is so, there is no absolute distinction between the types of transmission which we can use for sending a telegram from country to country and the types of transmission which at least are theoretically possible for transmitting a living organism such as a human being.

Let us then admit that the idea that one might conceivably travel by telegraph, in addition to traveling by train or airplane, is not intrinsically absurd, far as it may be from realization. The difficulties are, of course, enormous. It is possible to evaluate something like the amount of significant information conveyed by all the genes in a germ cell, and thereby to determine the amount of hereditary information, as compared with learned information, that a human being possesses. In order for this message to be significant at all, it must convey at least as much information as an entire set of the *Encyclopedia Britannica*. In fact if we compare the number of asymmetric carbon atoms in all the molecules of a germ cell with the number of dots and dashes needed to code the *Encyclopedia Britannica*, we find that they constitute an even more enormous message; and this is still more impressive when we realize what the conditions for telegraphic transmission of such a message must be. Any scanning of the human organism must be a probe going through all its parts, and will, accordingly, tend to destroy the tissue on its way. To hold an organism stable while part of it is being slowly destroyed, with the intention of re-creating it out of other material elsewhere, involves a lowering of its degree of activity, which in most cases would destroy life in the tissue.

In other words, the fact that we cannot telegraph the

pattern of a man from one place to another seems to be due to technical difficulties, and in particular, to the difficulty of keeping an organism in being during such a radical reconstruction. The idea itself is highly plausible. As for the problem of the radical reconstruction of the living organism, it would be hard to find any such reconstruction much more radical than that of a butterfly during its period as a pupa.

I have stated these things, not because I want to write a science fiction story concerning itself with the possibility of telegraphing a man, but because it may help us understand that the fundamental idea of communication is that of the transmission of messages, and that the bodily transmission of matter and messages is only one conceivable way of attaining that end. It will be well to reconsider Kipling's test of the importance of traffic in the modern world from the point of view of a traffic which is overwhelmingly not so much the transmission of human bodies as the transmission of human information.

VI

LAW AND COMMUNICATION

Law may be defined as the ethical control applied to communication, and to language as a form of communication, especially when this normative aspect is under the control of some authority sufficiently strong to give its decisions an effective social sanction. It is the process of adjusting the "couplings" connecting the behavior of different individuals in such a way that what we call justice may be accomplished, and disputes may be avoided, or at least adjudicated. Thus the theory and practice of the law involves two sets of problems: those of its general purpose, of its conception of justice; and those of the technique by which these concepts of justice can be made effective.

Empirically, the concepts of justice which men have maintained throughout history are as varied as the religions of the world, or the cultures recognized by anthropologists. I doubt if it is possible to justify them by any higher sanction than our moral code itself, which is indeed only another name for our conception of justice. As a participant in a liberal outlook which has its main roots in the Western tradition, but which has extended itself to those Eastern countries which have a strong intellectual-moral tradition, and has indeed borrowed deeply from them, I can only state what I myself and those about me consider necessary for the existence of justice. The best words to express these requirements are those of the French Revolution: *Liberté, Egalité, Fraternité.* These mean: the liberty of each human being to develop in his freedom the full measure of the human possibilities embodied in him; the equality by which what is just for A and B remains just when the positions of A and B are interchanged; and a good will between man and man that knows no limits short of those of humanity itself. These great principles of justice mean and demand that no person, by virtue of the personal strength of his position, shall enforce a sharp bargain by duress. What compulsion the very existence of the community and the state may demand must be exercised in such a way as to produce no unnecessary infringement of freedom.

But not even the greatest human decency and liberalism will, in itself, assure a fair and administrable legal code. Besides the general principles of justice, the law must be so clear and reproducible that the individual citizen can assess his rights and duties in advance, even where they appear to conflict with those of others. He must be able to ascertain with a reasonable certainty what view a judge or a jury will take of his position. If he cannot do this, the legal code, no matter how well intended, will not enable him to lead a life free from litigation and confusion.

Let us look at the matter from the simplest point of view—that of the contract. Here A takes on a responsibility of performing a certain service which in general will be advantageous to B; whereas B assumes in return the responsibility of performing a service or making a payment advantageous to A. If it is unambiguously clear what each task and each payment is to be, and if one of the parties does not invoke methods of imposing his will on the other party which are foreign to the contract itself, then the determination of whether the bargain is equitable may safely be left to the judgment of the two contracting parties. If it is manifestly inequitable, at least one of the contracting parties may be supposed to be in the position of being able to reject the bargain altogether. What, however, they cannot be expected to settle with any justice among themselves is the meaning of the bargain if the terms employed have no established significance, or if the significance varies from court to court. Thus it is the first duty of the law to see that the obligations and rights given to an individual in a certain stated situation be unambiguous. Moreover, there should be a body of legal interpretation which is as far as possible independent of the will and the interpretation of the particular authorities consulted. Reproducibility is prior to equity, for without it there can be no equity.

It appears from this why precedent has a very important theoretical weight in most legal systems, and why in all legal systems it has an important practical weight. There are those legal systems which purport to be based on certain abstract principles of justice. The Roman law and its descendants, which indeed constitute the greater part of the law of the European continent, belong to this class. There are other systems like that of the English law, in which it is openly stated that precedent is the main basis of legal thought. In either case, no new legal term has a completely secure meaning until it and its limitations have been determined in practice; and this is a matter of precedent. To fly in the face of a decision which has been made in an already existing case is to attack the uniqueness of the interpretation of legal language and is *ipso facto* a cause of indeterminateness and very probably of a consequent injustice. Every case decided should advance the definition of the legal terms involved in a manner consistent with past decisions, and it should lead naturally on to new ones. Every piece of phraseology should be tested by the custom of the place and of the field of human activity to which it is relevant. The judges, those to whom is confided the task of the interpretation of the law, should perform their function in such a spirit that if Judge A is replaced by Judge B, the exchange cannot be expected to make a material change in the court's interpretation of customs and of statutes. This naturally must remain to some extent an ideal rather than a *fait accompli;* but unless we are close followers of these ideals, we shall have chaos, and what is worse, a no-man's land in which dishonest men prey on the differences in possible interpretation of the statutes.

All of this is very obvious in the matter of contracts; but in fact it extends quite far into other branches of the law, and particularly of the civil law. Let me give an example. A, because of the carelessness of an employee B, damages a piece of property belonging to C. Who is to take the loss, and in what proportion? If these matters are known equally in advance to everybody, then it is possible for the person normally taking the greatest risk to charge a greater price for his undertakings and thus to insure himself. By these means he may cancel some considerable part of his disadvantage. The general effect of this is to spread the loss over the community, so that no man's share of it shall be ruinous. Thus the law of torts tends to partake somewhat of the same nature as the law of contracts. Any legal responsibility which involves exorbitant possibilities of loss will generally make the person incurring the loss pass his risk on to the community at large in the form of an increased price for his goods, or increased fees for his services. Here, as well as in the case of contracts, unambiguity, precedent, and a good clear tradition of interpretation are worth more than a theoretical equity, particularly in the assessment of responsibilities.

There are, of course, exceptions to these statements. For example, the old law of imprisonment for debt was inequitable in that it put the individual responsible for paying the debt in exactly that position in which he was least capable of acquiring the means to pay. There are many laws at present which are inequitable, because, for example, they assume a freedom of choice on the part of one party which under existing social circumstances is not there. What has been said about imprisonment for debt is equally valid in the case of peonage, and of many other similarly abused social customs.

If we are to carry out a philosophy of liberty, equality, and fraternity, then in addition to the demand that legal responsibility should be unambiguous, we must add the demand that it should not be of such a nature that one party acts under duress, leaving the other free. The history of our dealings with the Indians is full of instances in point, both for the dangers of duress and the dangers of ambiguity. From the very early times of the colonies, the Indians had neither the bulk of population nor the equality of arms to meet the whites on a fair basis, especially when the so-called land treaties between the whites and the Indians were being negotiated. Besides this gross injustice, there was a semantic injustice, which was perhaps even greater. The Indians as a hunting people had no idea of land as private property. For them there was no such ownership as ownership in fee simple, though they did have the notion of hunting rights over specific territories. In their treaties with the settlers, what they wished to convey were hunting rights, and generally only concomitant hunting rights over certain regions. On the other hand, the whites believed, if we are to give their conduct the most favorable interpretation that can be assigned to it, that the Indians were conveying to them a title to ownership in fee simple. Under these circumstances, not even a semblance of justice was possible, nor did it exist.

Where the law of Western countries is at present least satisfactory is on the criminal side. Law seems to consider punishment, now as a threat to discourage other possible criminals, now as a ritual act of expiation on the part of the guilty man, now as a device for removing him from society and for protecting the latter from the danger of repeated misconduct, and now as an agency for the social and the moral reform of the individual. These are four different tasks, to be accomplished by four different methods; and unless we know an accurate way of proportioning them, our whole attitude to the criminal will be at cross-purposes. At present, the criminal law speaks now in one language, and now in another. Until we in the community have made up our minds that what we really want is expiation, or removal, or reform, or the discouragement of potential criminals, we shall get none of these, but only a confusion in which crime breeds more crime. Any code which is made, one-fourth on the eighteenth-century British prejudice in favor of hanging, one-fourth on the removal of the criminal from society, one-fourth on a halfhearted policy of reform, and one-fourth on the policy of hanging up a dead crow to scare away the rest, is going to get us precisely nowhere.

Let us put it this way: the first duty of the law, whatever the second and third ones are, is to know what it wants. The first duty of the legislator or the judge is to make clear, unambiguous statements, which not only experts, but the common man of the times will interpret in one way and in one way only. The technique of the interpretation of past judgments must be such that a lawyer should know, not only what a court has said, but even with high probability what the court is going to say. Thus the problems of law may be considered communicative and cybernetic—that is, they are problems of orderly and repeatable control of certain critical situations.

There are vast fields of law where there is no satisfactory semantic agreement between what the law intends to say, and the actual situation that it contemplates. Whenever such a theoretical agreement fails to exist, we shall have the same sort of no-man's land that faces us when we have two currency systems without an accepted basis of exchange. In the zone of unconformity between one court and another or one coinage and another, there is always a refuge for the dishonest middleman, who will accept payment neither financially nor morally except in the system most favorable to him, and will give it only in the system in which he sacrifices least. The greatest opportunity of the criminal in the modern community lies in this position as a dishonest broker in the interstices of the law. I have pointed out in an earlier chapter that noise, regarded as a confusing factor in human communications, is damaging but not consciously malicious. This is true as far as scientific communication goes, and to a large extent in ordinary conversation between two people. It is most emphatically not true in language as it is used in the law courts.

The whole nature of our legal system is that of a conflict. It is a conversation in which at least three parties take part—let us say, in a civil case, the plaintiff, the defendant, and the legal system as represented by judge and jury. It is a game in the full Von Neumann sense; a game in which the litigants try by methods which are limited by the code of law to obtain the judge and the jury as their partners. In such a game the opposing lawyer, unlike nature itself, can and deliberately does try to introduce confusion into the messages of the side he is opposing. He tries to reduce their statements to nonsense, and he deliberately jams the messages between his antagonist and the judge and jury. In this jamming, it is inevitable that bluff should occasionally be at a premium. Here we do not need to take the Erle Stanley Gardner detective stories at their face value as descriptions of legal procedure to see that there are occasions in litigation where bluff or the sending of messages with a deliberate purpose of concealing the strategy of the sender is not only permitted but encouraged.

VII

COMMUNICATION, SECRECY,
AND SOCIAL POLICY

In the world of affairs, the last few years have been characterized by two opposite, even contradictory, trends. On the one hand, we have a network of communication, intranational and international, more complete than history has ever before seen. On the other hand, under the impetus of Senator McCarthy and his imitators, the blind and excessive classification of military information, and the recent attacks on the State Department, we are approaching a secretive frame of mind paralleled in history only in the Venice of the Renaissance.

There the extraordinarily precise news-gathering services of the Venetian ambassadors (which form one of our chief sources of European history) accompanied a national jealousy of secrets, exaggerated to such an extent that the state ordered the private assassination of emigrant artisans, to maintain the monopoly of certain chosen arts and crafts. The modern game of cops and robbers which seems to characterize both Russia and the United States, the two principal contestants for world power of this century, suggests the old Italian cloak-and-dagger melodrama played on a much larger stage.

The Italy of the Renaissance was also the scene of the birth-pangs of modern science. However, the science of the present day is a much larger undertaking than that of Renaissance Italy. It should be possible to examine all the elements of information and secrecy in the modern world with a somewhat greater maturity and objectivity than belong to the thought of the times of Machiavelli. This is particularly so in view of the fact that, as we have seen, the study of communication has now reached a degree of independence and authority making it a science in its own right. What does modern science have to say concerning the status and functions of communication and secrecy?

I am writing this book primarily for Americans in whose environment questions of information will be evaluated according to a standard American criterion: a thing is valuable as a commodity for what it will bring in the open market. This is the official doctrine of an orthodoxy which it is becoming more and more perilous for a resident of the United States to question. It is perhaps worth while to point out that it does not represent a universal basis of human values: that it corresponds neither to the doctrine of the Church, which seeks for the salvation of the human soul, nor to that of Marxism, which values a society for its realization of certain specific ideals of human well-being. The fate of information in the typically American world is to become something which can be bought or sold.

It is not my business to cavil whether this mercantile attitude is moral or immoral, crass or subtle. It is my business to show that it leads to the misunderstanding and the mistreatment of information and its associated concepts. I shall take this up in several fields, beginning with that of patent law.

The letters patent granting to an inventor a limited monopoly over the subject matter of his invention are for him what a charter is to a chartered company. Behind our patent law and our patent policy is an implicit philosophy of private property and of the rights thereto. This philosophy represented a fairly close approximation to the actual situation in the now ending period when inventions were generally made in the shop by skilled handicraftsmen. It does not represent even a passable picture of the inventions of the present day.

The standard philosophy of the patent office presupposes that by a system of trial and error, implying what is generally called mechanical ingenuity, a craftsman has advanced from a given technique to a further stage, embodied in a specific apparatus. The law distinguishes the ingenuity which is necessary to make this new combination from the other sort of ingenuity which is necessary to find out scientific facts about the world. This second sort of ingenuity is labeled the *discovery of a law of nature;* and in the United States, as well as in many other countries with similar industrial practices, the legal code denies to the discoverer any property rights in a law of nature which he may have discovered. It will be seen that at one time this distinction was fairly practical, for the shop inventor has one sort of tradition and background, and the man of science has a totally different one.

The Daniel Doyce of Dickens' *Little Dorrit,* is clearly not to be mistaken for the members of the Mudfog Association which Dickens treats elsewhere. The first, Dickens glorifies as the common sense craftsman, with the broad thumb of the hand worker, and the honesty of the man who is always facing facts; whereas the Mudfog Association is nothing but a derogatory alias for the British Association for the Advancement of Science in its early days. Dickens reviles the latter as an assemblage of chimerical and useless dreamers, in language which Swift would not have found inadequate to describe the projectors of Laputa.

Now a modern research laboratory such as that of the Bell Telephone Company, while it retains Doyce's practicality, actually consists of the great-grandchildren of the Mudfog Association. If we take Faraday as an outstanding yet typical member of the early British Association for the Advancement of Science, the chain to the research men of the Bell Telephone Laboratories of the present day is complete, by way of Maxwell and Heaviside, to Campbell and Shannon.

In the early days of modern invention, science was far ahead of the workman. The locksmith set the level of mechanical competence. A piston was considered to fit an engine-cylinder when, according to Watt, a thin sixpence could just be slipped between the two. Steel was a craftsman's product, for swords and armor; iron was the stringy, slag-filled product of the puddler. Daniel Doyce had a long way indeed to go before so practical a scientist as Faraday could begin to supplant him. It is not strange that the policy of Great Britain, even when expressed through such a purblind organ as Dickens' Circumlocution Office, was directed toward Doyce as the true inventor, rather than to the gentlemen of the Mudfog Society. The Barnacle family of hereditary bureaucrats might wear Doyce to a shadow, before they ceased to refer him from office to office, but they secretly feared him, as the representative of the new industrialism which was displacing them. They neither feared, respected, nor understood the gentlemen of the Mudfog Association.

In the United States, Edison represents the precise transition between the Doyces and the men of the Mudfog Association. He was himself very much of a Doyce, and was even more desirous of appearing to be one. Nevertheless, he chose much of his staff from the Mudfog camp. His greatest invention was that of the industrial research laboratory, turning out inventions as a business. The General Electric Company, the Westinghouse interests, and the Bell Telephone Laboratories followed in his footsteps, employing scientists by hundreds where Edison employed them by tens. Invention came to mean, not the gadget-insight of a shop-worker, but the result of a careful, comprehensive search by a team of competent scientists.

At present, the invention is losing its identity as a

commodity in the face of the general intellectual structure of emergent inventions. What makes a thing a good commodity? Essentially, that it can pass from hand to hand with the substantial retention of its value, and that the pieces of this commodity should combine additively in the same way as the money paid for them. The power to conserve itself is a very convenient property for a good commodity to have. For example, a given amount of electrical energy, except for minute losses, is the same at both ends of a transmission line, and the problem of putting a fair price on electric energy in kilowatt-hours is not too difficult. A similar situation applies to the law of the conservation of matter. Our ordinary standards of value are quantities of gold, which is a particularly stable sort of matter.

Information, on the other hand, cannot be conserved as easily, for as we have already seen the amount of information communicated is related to the non-additive quantity known as entropy and differs from it by its algebraic sign and a possible numerical factor. Just as entropy tends to increase spontaneously in a closed system, so information tends to decrease; just as entropy is a measure of disorder, so information is a measure of order. Information and entropy are not conserved, and are equally unsuited to being commodities.

In considering information or order from the economic point of view, let us take as an example a piece of gold jewelry. The value is composed of two parts: the value of the gold, and that of the "façon," or workmanship. When an old piece of jewelry is taken to the pawnbroker or the appraiser, the firm value of the piece is that of the gold only. Whether a further allowance is made for the façon or not depends on many factors, such as the persistence of the seller, the style in favor when it was made, the purely artistic craftsmanship, the historical value of the piece for museum purposes, and the resistance of the buyer.

Many a fortune has been lost by ignoring the difference between these two types of values, that of the gold and that of the façon. The stamp market, the rare-book market, the market for Sandwich glass and for Duncan Phyfe furniture are all artificial, in the sense that in addition to the real pleasure which the possession of such an object gives to its owner, much of the value of the façon pertains not only to the rarity of the object itself, but to the momentary existence of an active group of buyers competing for it. A depression, which limits the group of possible buyers, may divide it by a factor of four or five, and a great treasure vanishes into nothing just for want of a competitive purchaser. Let another new popular craze supplant the old in the attention of the prospective collectors, and again the bottom may drop out of the market. There is no permanent common denominator of collectors' taste, at least until one approaches the highest level of aesthetic value. Even then the prices paid for great paintings are colossal reflections of the desire of the purchaser for the reputation of wealth and connoisseurdom.

The problem of the work of art as a commodity raises a large number of questions important in the theory of information. In the first place, except in the case of the narrowest sort of collector who keeps all his possessions under permanent lock and key, the physical possession of a work of art is neither sufficient nor necessary for the benefits of appreciation which it conveys. Indeed, there are certain sorts of works of art which are essentially public rather than private in their appeal, and concerning which the problem of possession is almost irrelevant. A great fresco is scarcely a negotiable document, nor for that matter is the building on whose walls it is placed. Whoever is technically the possessor of such works of art must share them at least with the limited public that frequents the buildings, and very often with the world at large. He cannot place them

in a fireproof cabinet and gloat over them at a small
dinner for a few connoisseurs, nor shut them up altogether as private possessions. There are very few fres-
coes which are given the adventitious privacy of the
one by Siqueiros which adorns a large wall of the Mexican jail where he served a sentence for a political
offense.

So much for the mere physical possession of a work
of art. The problems of property in art lie much deeper.
Let us consider the matter of the reproduction of artistic works. It is without a doubt true that the finest
flower of artistic appreciation is only possible with originals, but it is equally true that a broad and culti-
vated taste may be built up by a man who has never
seen an original of a great work, and that by far the
greater part of the aesthetic appeal of an artistic creation is transmitted in competent reproductions. The
case of music is similar. While the hearer gains something very important in the appreciation of a musical
composition if he is physically present at the performance, nevertheless his preparation for an understanding of this performance will be so greatly enhanced by
hearing good records of the composition that it is hard
to say which of the two is the larger experience.

From the standpoint of property, reproduction-
rights are covered by our copyright law. There are
other rights which no copyright law can cover, which
almost equally raise the question of the capacity of
any man to own an artistic creation in an effective
sense. Here the problem of the nature of genuine originality arises. For example, during the period of the
high Renaissance, the discovery by the artists of ge-
ometric perspective was new, and an artist was able
to give great pleasure by the skillful exploitation of this
element in the world about him. Dürer, Da Vinci, and
their contemporaries exemplify the interest which the
leading artistic minds of the time found in this new
device. As the art of perspective is one which, once

mastered, rapidly loses its interest, the same thing that was great in the hands of its originators is now at the disposal of every sentimental commercial artist who designs trade calendars.

What has been said before may not be worth saying again; and the informative value of a painting or a piece of literature cannot be judged without knowing what it contains that is not easily available to the public in contemporary or earlier works. It is only independent information which is even approximately additive. The derivative information of the second-rate copyist is far from independent of what has gone before. Thus the conventional love story, the conventional detective story, the average acceptable success tale of the slicks, all are subject to the letter but not the spirit of the law of copyright. There is no form of copyright law that prevents a movie success from being followed by a stream of inferior pictures exploiting the second and third layers of the public's interest in the same emotional situation. Neither is there a way of copyrighting a new mathematical idea, or a new theory such as that of natural selection, or anything except the identical reproduction of the same idea in the same words.

I repeat, the prevalence of clichés is no accident, but inherent in the nature of information. Property rights in information suffer from the necessary disadvantage that a piece of information, in order to contribute to the general information of the community, must say something substantially different from the community's previous common stock of information. Even in the great classics of literature and art, much of the obvious informative value has gone out of them, merely by the fact that the public has become acquainted with their contents. Schoolboys do not like Shakespeare, because he seems to them nothing but a mass of familiar quotations. It is only when the study of such an author has penetrated to a layer deeper than that which has been absorbed into the superficial clichés of the time, that we can re-establish with him an informative *rapport,* and give him a new and fresh literary value.

It is interesting from this point of view that there are authors and painters who, by their wide exploration of the aesthetic and intellectual avenues open to a given age, have an almost destructive influence on their contemporaries and successors for many years. A painter like Picasso, who runs through many periods and phases, ends up by saying all those things which are on the tip of the tongue of the age to say, and finally sterilizes the originality of his contemporaries and juniors.

The intrinsic limitations of the commodity nature of communication are hardly considered by the public at large. The man in the street considers that Maecenas had as his function the purchase and storage of works of art, rather than the encouragement of their creation by the artists of his own time. In a quite analogous way, he believes that it is possible to store up the military and scientific know-how of the nation in static libraries and laboratories, just as it is possible to store up the military weapons of the last war in the arsenals. Indeed, he goes further, and considers that information which has been developed in the laboratories of his own country is morally the property of that country; and that the use of this information by other nationalities not only may be the result of treason, but intrinsically partakes of the nature of theft. He cannot conceive of a piece of information without an owner.

The idea that information can be stored in a changing world without an overwhelming depreciation in its value is false. It is scarcely less false than the more plausible claim, that after a war we may take our existing weapons, fill their barrels with cylinder oil, and coat their outsides with sprayed rubber film, and let them statically await the next emergency. Now, in view of the changes in the technique of war, rifles store fairly well, tanks poorly, and battleships and submarines not at all. The fact is that the efficacy of a weapon depends on precisely what other weapons there are to meet it at a given time, and on the whole idea of war at that time. This results—as has been proved more than once —in the existence of excessive stockpiles of stored weapons which are likely to stereotype the military policy in a wrong form, so that there is a very appreciable advantage to approaching a new emergency with the freedom of choosing exactly the right tools to meet it.

On another level, that of economics, this is conspicuously true, as the British example shows. England was the first country to go through a full-scale industrial revolution; and from this early age it inherited the narrow gauge of its railways, the heavy investment of its cotton mills in obsolete equipment, and the limitations of its social system, which have made the cumulative needs of the present day into an overwhelming emergency, only to be met by what amounts to a social and industrial revolution. All this is taking place while the newest countries to industrialize are able to enjoy the latest, most economical equipment; are able to construct an adequate system of railroads to carry their goods on economically-sized cars; and in general, are able to live in the present day rather than in that of a century ago.

What is true of England is true of New England, which has discovered that it is often a far more expensive matter to modernize an industry than to scrap it and to start somewhere else. Quite apart from the difficulties of having a relatively strict industrial law and an advanced labor policy, one of the chief reasons that New England is being deserted by the textile mills is that, frankly, they prefer not to be hampered by a century of traditions. Thus, even in the most material field, production and security are in the long run matters of continued invention and development.

Information is more a matter of process than of storage. That country will have the greatest security whose informational and scientific situation is adequate to meet the demands that may be put on it—the country in which it is fully realized that information is important as a stage in the continuous process by which we observe the outer world, and act effectively upon it. In other words, no amount of scientific research, carefully recorded in books and papers, and then put into our libraries with labels of secrecy, will be adequate to protect us for any length of time in a world where the effective level of information is perpetually advancing. There is no Maginot Line of the brain.

I repeat, to be alive is to participate in a continuous stream of influences from the outer world and acts on the outer world, in which we are merely the transitional stage. In the figurative sense, to be alive to what is happening in the world, means to participate in a continual development of knowledge and its unhampered exchange. In anything like a normal situation, it is both far more difficult and far more important for us to ensure that we have such an adequate knowledge than to ensure that some possible enemy does *not* have it. The whole arrangement of a military research laboratory is along lines hostile to our own optimum use and development of information.

During the last war an integral equation of a type which I have been to some extent responsible for solving arose, not only in my own work, but in at least two totally unrelated projects. In one of these I was aware that it was bound to arise; and in the other a very slight amount of consultation should have made me so aware. As these three employments of the same idea belonged to three totally different military projects of totally different levels of secrecy and in diverse places, there was no way by which the information of any one of them could penetrate through to the others. The result was that it took the equivalent of three independent discoveries to make the results accessible in all three fields. The delay thus created was a matter of from some six months to a year, and probably considerably more. From the standpoint of money, which of course is less important in war, it amounted to a large number of man-years at a very expensive level. It would take a considerable valuable employment of this work by an enemy to be as disadvantageous as the need for reproducing all the work on our part. Remember that an enemy unable to participate in that residual discussion which takes place quite illegally, even under our setup of secrecy, would not have been in the position to evaluate and use our results.

The matter of time is essential in all estimates of the value of information. A code or cipher, for example, which will cover any considerable amount of material at high-secrecy level is not only a lock which is hard to force, but also one which takes a considerable time to open legitimately. Tactical information which is useful in the combat of small units will almost certainly be obsolete in an hour or two. It is a matter of very little importance whether it can be broken in three hours; but it is of great importance that an officer receiving the message should be able to read it in something like two minutes. On the other hand, the larger plan of battle is too important a matter to entrust to this limited degree of security. Nevertheless, if it took a whole day for an officer receiving this plan to disentangle it, the delay might well be more serious than any leak. The codes and ciphers for a whole campaign or for a diplomatic policy might and should be still less easy to penetrate; but there are none which cannot be penetrated in any finite period, and which at the same time can carry a significant amount of information rather than a small set of disconnected individual decisions.

The ordinary way of breaking a cipher is to find an example of the use of this cipher sufficiently long so that the pattern of encodement becomes obvious to the skilled investigator. In general, there must be at least a minimum degree of repetition of patterns, without which the very short passages lacking repetition cannot be deciphered. However, when a number of passages are enciphered in a type of cipher which is common to the whole set, even though the detailed encipherment varies, there may be enough in common between the different passages to lead to a breaking, first of the general type of cipher, and then of the particular ciphers used.

Probably much of the greatest ingenuity which has been shown in the breaking of ciphers appears not in the annals of the various secret services, but in the work of the epigrapher. We all know how the Rosetta Stone was decoded through an interpretation of certain characters in the Egyptian version, which turned out to be the names of the Ptolemies. There is however one act of decoding which is greater still. This greatest single example of the art of decoding is the decoding of the secrets of nature itself and is the province of the scientist.

Scientific discovery consists in the interpretation for our own convenience of a system of existence which has been made with no eye to our convenience at all. The result is that the last thing in the world suitable for the protection of secrecy and elaborate code system is a law of nature. Besides the possibility of breaking the secrecy by a direct attack on the human or documentary vehicles of this secrecy, there is always the possibility of attacking the code upstream of all these. It is perhaps impossible to devise any secondary code as hard to break as the natural code of the atomic nucleus.

In the problem of decoding, the most important information which we can possess is the knowledge that the message which we are reading is not gibberish. A common method of disconcerting codebreakers is to mix in with the legitimate message a message that cannot be decoded; a non-significant message, a mere assemblage of characters. In a similar way, when we consider a problem of nature such as that of atomic reactions and atomic explosives, the largest single item of information which we can make public is that they exist. Once a scientist attacks a problem which he knows to have an answer, his entire attitude is changed. He is already some fifty per cent of his way toward that answer.

In view of this, it is perfectly fair to say that the one secret concerning the atomic bomb which might have been kept and which was given to the public and to all potential enemies without the least inhibition, was that of the possibility on its construction. Take a problem of this importance and assure the scientific world that it has an answer; then both the intellectual ability of the scientists and the existing laboratory facilities are so widely distributed that the quasi-independent realization of the task will be a matter of merely a few years anywhere in the world.

There is at present a touching belief in this country that we are the sole possessors of a certain technique called "know-how," which secures for us not only priority on all engineering and scientific developments and all major inventions, but, as we have said, the moral right to that priority. Certainly, this "know-how" has nothing to do with the national origins of those who have worked on such problems as that of the atomic bomb. It would have been impossible throughout most of history to secure the combined services of such scientists as the Dane, Bohr; the Italian, Fermi; the Hungarian, Szilard; and many others involved in the project. What made it possible was the extreme consciousness of emergency and the sense of universal affront excited by the Nazi threat. Something more than inflated propaganda will be necessary to hold such a group together over the long period of rearmament to which we have often seemed to be committed by the policy of the State Department.

Without any doubt, we possess the world's most highly developed technique of combining the efforts of large numbers of scientists and large quantities of money toward the realization of a single project. This should not lead us to any undue complacency concerning our scientific position, for it is equally clear that we are bringing up a generation of young men who cannot think of any scientific project except in terms of large numbers of men and large quantities of money. The skill by which the French and English do great amounts of work with apparatus which an American high-school teacher would scorn as a casual stick-and-string job, is not to be found among any but a vanishingly small minority of our young men. The present vogue of the big laboratory is a new thing in science. There are those of us who wish to think that it may never last to be an old thing, for when the scientific ideas of this generation are exhausted, or at least reveal vastly diminishing returns on their intellectual investment, I do not foresee that the next generation will be able to furnish the colossal ideas on which colossal projects naturally rest.

A clear understanding of the notion of information as applied to scientific work will show that the simple coexistence of two items of information is of relatively small value, unless these two items can be effectively combined in some mind or organ which is able to fertilize one by means of the other. This is the very opposite of the organization in which each member travels a preassigned path, and in which the sentinels of science, when they come to the ends of their beats, present arms, do an about face, and march back in the direction from which they have come. There is a great fertilizing and revivifying value in the contact of two scientists with each other; but this can only come when at least one of the human beings representing the science has penetrated far enough across the frontier to be able to absorb the ideas of his neighbor into an effective plan of thinking. The natural vehicle for this type of organization is a plan in which the orbit of each scientist is assigned rather by the scope of his interests than as a predetermined beat.

Such loose human organizations do exist even in the United States; but at present they represent the result of the efforts of a few disinterested men, and not the planned frame into which we are being forced by those who imagine they know what is good for us. However, it will not do for the masses of our scientific population to blame their appointed and self-appointed betters for their futility, and for the dangers of the present day. It is the great public which is demanding the utmost of secrecy for modern science in all things which may touch its military uses. This demand for secrecy is scarcely more than the wish of a sick civilization not to learn of the progress of its own disease. So long as we can continue to pretend that all is right with the world, we plug up our ears against the sound of "Ancestral voices prophesying war."

In this new attitude of the masses at large to research, there is a revolution in science far beyond what the public realizes. Indeed the lords of the present science themselves do not foresee the full consequences of what is going on. In the past the direction of research had largely been left to the interest of the individual scholar and to the trend of the times. At present, there is a distinct attempt so to direct research in matters of public security that as far as possible, all significant avenues will be developed with the objective of securing an impenetrable stockade of scientific protection. Now, science is impersonal, and the result of a further pushing forward of the frontiers of science is not merely to show us many weapons which we may employ against possible enemies, but also many dangers of these weapons. These may be due to the fact that they either are precisely those weapons which are more effectively employable against us than against any enemy of ours, or are dangers, such as that of radio-active poisoning, which are inherent in our very use of such a weapon as the atomic bomb. The hurrying up of the pace of science, owing to our active simultaneous search for all means of attacking our enemies and of protecting ourselves, leads to ever-increasing demands for new research. For example, the concentrated effort of Oak Ridge and Los Alamos in time of war has made the question of the protection of the people of the United States, not only from the possible enemies employing an atomic bomb, but from the atomic radiation of our new industry, a thing which concerns us *now*. Had the war not occurred, these *perils* would probably not have concerned us for twenty years. In our present militaristic frame of mind, this has forced on us the problem of possible countermeasures to a new employment of these agencies on the part of an enemy. This enemy may be Russia at the present moment, but it is even more the reflection of ourselves in a mirage. To defend ourselves against this phantom, we must look to new scientific measures, each more terrible than the last. There is no end to this vast apocalyptic spiral.

We have already depicted litigation as a true game in which the antagonists can and are forced to use the full resources of bluff and thus each to develop a policy which may have to allow for the other player's playing the best possible game. What is true in the limited war of the court is also true in the war to the death of international relations, whether it takes the bloody form of shooting, or the suaver form of diplomacy.

The whole technique of secrecy, message jamming, and bluff, is concerned with insuring that one's own side can make use of the forces and agencies of communication more effectively than the other side. In this combative use of information it is quite as important to keep one's own message channels open as to obstruct the other side in the use of the channels available to it.

An over-all policy in matters of secrecy almost always must involve the consideration of many more things than secrecy itself.

We are in the position of the man who has only two ambitions in life. One is to invent the universal solvent which will dissolve any solid substance, and the second is to invent the universal container which will hold any liquid. Whatever this inventor does, he will be frustrated. Furthermore, as I have already said, no secret will ever be as safe when its protection is a matter of human integrity, as when it was dependent on the difficulties of scientific discovery itself.

I have already said the dissemination of any scientific secret whatever is merely a matter of time, that in this game a decade is a long time, and that in the long run, there is no distinction between arming ourselves and arming our enemies. Thus each terrifying discovery merely increases our subjection to the need of making a new discovery. Barring a new awareness on the part of our leaders, this is bound to go on and on, until the entire intellectual potential of the land is drained from any possible constructive application to the manifold needs of the race, old and new. The effect of these weapons must be to increase the entropy of this planet, until all distinctions of hot and cold, good and bad, man and matter have vanished in the formation of the white furnace of a new star.

Like so many Gadarene swine, we have taken unto us the devils of the age, and the compulsion of scientific warfare is driving us pell-mell, head over heels into the ocean of our own destruction. Or perhaps we may say that among the gentlemen who have made it their business to be our mentors, and who administer the new program of science, many are nothing more than apprentice sorcerers, fascinated with the incantation which starts a devilment that they are totally unable to stop. Even the new psychology of advertising and salesmanship becomes in their hands a way for obliterating the conscientious scruples of the working scientists, and for destroying such inhibitions as they may have against rowing into this maelstrom.

Let these wise men who have summoned a demoniac sanction for their own private purposes remember that in the natural course of events, a conscience which has been bought once will be bought twice. The loyalty to humanity which can be subverted by a skillful distribution of administrative sugar plums will be followed by a loyalty to official superiors lasting just so long as we have the bigger sugar plums to distribute. The day may well come when it constitutes the biggest potential threat to our own security. In that moment in which some other power, be it fascist or communist, is in the position to offer the greater rewards, our good friends who have rushed to our defense per account rendered will rush as quickly to our subjection and annihilation. May those who have summoned from the deep the spirits of atomic warfare remember that for their own sake, if not for ours, they must not wait beyond the first glimmerings of success on the part of our opponents to put to death those whom they have already corrupted!

VIII

ROLE OF THE INTELLECTUAL
AND THE SCIENTIST

This book argues that the integrity of the channels of internal communication is essential to the welfare of society. This internal communication is subject at the present time not only to the threats which it has faced at all times, but to certain new and especially serious problems which belong peculiarly to our age. One among these is the growing complexity and cost of communication.

A hundred and fifty years ago or even fifty years ago—it does not matter which—the world and America in particular were full of small journals and presses through which almost any man could obtain a hearing. The country editor was not as he is now limited to boiler plate and local gossip, but could and often did express his individual opinion, not only of local affairs but of world matters. At present this license to express oneself has become so expensive with the increasing cost of presses, paper, and syndicated services, that the newspaper business has come to be the art of saying less and less to more and more.

The movies may be quite inexpensive as far as concerns the cost of showing each show to each spectator, but they are so horribly expensive in the mass that few shows are worth the risk, unless their success is certain in advance. It is not the question whether a show may excite a great interest in a considerable number of people that interests the entrepreneur, but rather the question of whether it will be unacceptable to so few that he can count on selling it indiscriminately to movie theaters from coast to coast.

What I have said about the newspapers and the movies applies equally to the radio, to television, and even to bookselling. Thus we are in an age where the enormous per capita bulk of communication is met by an ever-thinning stream of total bulk of communication. More and more we must accept a standardized inoffensive and insignificant product which, like the white bread of the bakeries, is made rather for its keeping and selling properties than for its food value.

This is fundamentally an external handicap of modern communication, but it is paralleled by another which gnaws from within. This is the cancer of creative narrowness and feebleness.

In the old days, the young man who wished to enter the creative arts might either have plunged in directly or prepared himself by a general schooling, perhaps irrelevant to the specific tasks he finally undertook, but which was at least a searching discipline of his abilities and taste. Now the channels of apprenticeship are largely silted up. Our elementary and secondary schools are more interested in formal classroom discipline than in the intellectual discipline of learning something thoroughly, and a great deal of the serious preparation for a scientific or a literary course is relegated to some sort of graduate school or other.

Hollywood meanwhile has found that the very standardization of its product has interfered with the natural flow of acting talent from the legitimate stage. The repertory theaters had almost ceased to exist when some of them were reopened as Hollywood talent farms, and even these are dying on the vine. To a considerable extent our young would-be actors have learned their trade, not on the stage, but in university courses on acting. Our writers cannot get very far as young men in competition with syndicate material, and if they do not make a success the first try, they have no place to go but college courses which are supposed to teach them how to write. Thus the higher degrees, and above all the Ph.D., which have had a long existence as the legitimate preparation of the scientific specialist, are more and more serving as a model for intellectual training in all fields.

Properly speaking the artist, the writer, and the scientist should be moved by such an irresistible impulse to create that, even if they were not being paid for their work, they would be willing to pay to get the chance to do it. However, we are in a period in which forms have largely superseded educational content and one which is moving toward an ever-increasing thinness of educational content. It is now considered perhaps more a matter of social prestige to obtain a higher degree and follow what may be regarded as a cultural career, than a matter of any deep impulse.

In view of this great bulk of semi-mature apprentices who are being put on the market, the problem of giving them some colorable material to work on has assumed an overwhelming importance. Theoretically they should find their own material, but the big business of modern advanced education cannot be operated under this relatively low pressure. Thus the earlier stages of creative work, whether in the arts or in the sciences, which should properly be governed by a great desire on the part of the students to create something and to communicate it to the world at large, are now subject instead to the formal requirements of finding Ph.D. theses or similar apprentice media.

Some of my friends have even asserted that a Ph.D. thesis should be the greatest scientific work a man has ever done and perhaps ever will do, and should wait until he is thoroughly able to state his life work. I do not go along with this. I mean merely that if the thesis is not in fact such an overwhelming task, it should at least be in intention the gateway to vigorous creative work. Lord only knows that there are enough problems yet to be solved, books to be written, and music to be composed! Yet for all but a very few, the path to these lies through the performance of perfunctory tasks which in nine cases out of ten have no compelling reason to be performed. Heaven save us from the first novels which are written because a young man desires the prestige of being a novelist rather than because he has something to say! Heaven save us likewise from the mathematical papers which are correct and elegant but without body or spirit. Heaven save us above all from the snobbery which not only admits the possibility of this thin and perfunctory work, but which cries out in a spirit of shrinking arrogance against the competition of vigor and ideas, wherever these may be found!

In other words, when there is communication without need for communication, merely so that someone may earn the social and intellectual prestige of becoming a priest of communication, the quality and communicative value of the message drop like a plummet. It is as if a machine should be made from the Rube Goldberg point of view, to show just what recondite ends may be served by an apparatus apparently quite unsuitable for them, rather than to do something. In the arts, the desire to find new things to say and new ways of saying them is the source of all life and interest. Yet every day we meet with examples of painting where, for instance, the artist has bound himself from the new canons of the abstract, and has displayed no intention to use these canons to display an interesting and novel form of beauty, to pursue the uphill fight against the prevailing tendency toward the commonplace and the banal. Not all the artistic pedants are academicians. There are pedantic *avantgardistes*. No school has a monoply on beauty. Beauty, like order, occurs in many places in this world, but only as a local and temporary fight against the Niagara of increasing entropy.

I speak here with feeling which is more intense as far as concerns the scientific artist than the conventional artist, because it is in science that I have first chosen to say something. What sometimes enrages me and always disappoints and grieves me is the preference of great schools of learning for the derivative as opposed to the original, for the conventional and thin which can be duplicated in many copies rather than the new and powerful, and for arid correctness and limitation of scope and method rather than for universal newness and beauty, wherever it may be seen. Moreover, I protest, not only as I have already done against the cutting off of intellectual originality by the difficulties of the means of communication in the modern world, but even more against the ax which has been put to the root of originality because the people who have elected communication as a career so often have nothing more to communicate.

IX

THE FIRST AND THE SECOND
INDUSTRIAL REVOLUTION

The preceding chapters of this book dealt primarily with the study of man as a communicative organism. However, as we have already seen, the machine may also be a communicative organism. In this chapter, I shall discuss that field in which the communicative characters of man and of the machine impinge upon one another, and I shall try to ascertain what the direction of the development of the machine will be, and what we may expect of its impact on human society.

Once before in history the machine had impinged upon human culture with an effect of the greatest moment. This previous impact is known as the Industrial Revolution, and it concerned the machine purely as an alternative to human muscle. In order to study the present crisis, which we shall term The Second Industrial Revolution, it is perhaps wise to discuss the history of the earlier crisis as something of a model.

The first industrial revolution had its roots in the intellectual ferment of the eighteenth century, which found the scientific techniques of Newton and Huygens already well developed, but with applications which had yet scarcely transcended astronomy. It had, however, become manifest to all intelligent scientists that the new techniques were gong to have a profound effect on the other sciences. The first fields to show the impact of the Newtonian era were those of navigation and of clockmaking.

Navigation is an art which dates to ancient times, but

it had one conspicuous weakness until the seventeen-thirties. The problem of determining latitude had always been easy, even in the days of the Greeks. It was simply a matter of determining the angular height of the celestial pole. This may be done roughly by taking the pole star as the actual pole of the heavens, or it may be done very precisely by further refinements which locate the center of the apparent circular path of the pole star. On the other hand, the problem of longitudes is always more difficult. Short of a geodetic survey, it can be solved only by a comparison of local time with some standard time such as that of Greenwich. In order to do this, we must either carry the Greenwich time with us on a chronometer or we must find some heavenly clock other than the sun to take the place of a chronometer.

Before either of these two methods had become available for the practical navigator, he was very considerably hampered in his techniques of navigation. He was accustomed to sail along the coast until he reached the latitude he wanted. Then he would strike out on an east or west course, along a parallel of latitude, until he made a landfall. Except by an approximate dead-reckoning, he could not tell how far he was along the course, yet it was a matter of great importance to him that he should not come unawares onto a dangerous coast. Having made his landfall, he sailed along the coast until he came to his destination. It will be seen that under these circumstances every voyage was very much of an adventure. Nevertheless, this was the pattern of voyages for many centuries. It can be recognized in the course taken by Columbus, in that of the Silver Fleet, and that of the Acapulco galleons.

This slow and risky procedure was not satisfactory to the admiralties of the eighteenth century. In the first place, the overseas interests of England and France, unlike those of Spain, lay in high latitudes, where the advantage of a direct great-circle course over an east-and-west course is most conspicuous. In the second place, there was a great competition between the two northern powers for the supremacy of the seas, and the advantage of a better navigation was a serious one. It is not a surprise that both governments offered large rewards for an accurate technique of finding longitudes.

The history of these prize contests is complicated and not too edifying. More than one able man was deprived of his rightful triumph, and went bankrupt. In the end, these prizes were awarded in both countries for two very different achievements. One was the design of an accurate ship's chronometer—that is, of a clock sufficiently well constructed and compensated to be able to keep the time within a few seconds over a voyage in which it was subject to the continual violent motion of the ship. The other was the construction of good mathematical tables of the motion of the moon, which enabled the navigator to use that body as the clock with which to check the apparent motion of the sun. These two methods have dominated all navigation until the recent development of radio and radar techniques.

Accordingly, the advance guard of the craftsmen of the industrial revolution consisted on the one hand of clockmakers, who used the new mathematics of Newton in the design of their pendulums and their balance wheels; and on the other hand, of optical-instrument makers, with their sextants and their telescopes. The two trades had very much in common. They both demanded the construction of accurate circles and accurate straight lines, and the graduation of these in degrees or in inches. Their tools were the lathe and the dividing engine. These machine tools for delicate work are the ancestors of our present machine-tool industry.

It is an interesting reflection that every tool has a genealogy, and that it is descended from the tools by which it has itself been constructed. The clockmakers' lathes of the eighteenth century have led through a clear historical chain of intermediate tools to the great turret lathes of the present day. The series of intervening steps might conceivably have been foreshortened somewhat, but it has necessarily had a certain minimum length. It is clearly impossible in constructing a great turret lathe to depend on the unaided human hand for the pouring of the metal, for the placing of the castings on the instruments to machine them, and above all for the power needed in the task of machining them. These must be done through machines that have themselves been manufactured by other machines, and it is only through many stages of this that one reaches back to the original hand- or foot-lathes of the eighteenth century.

It is thus entirely natural that those who were to develop new inventions were either clockmakers or scientific-instrument makers themselves, or called on people of these crafts to help them. For instance, Watt was a scientific-instrument maker. To show how even a man like Watt had to bide his time before he could extend the precision of clockmaking techniques to larger undertakings, we must remember, as I have said earlier, that his standard of the fit of a piston in a cylinder was that it should be barely possible to insert and move a thin sixpence between them.

We must thus consider navigation and the instruments necessary for it as the locus of an industrial revolution before the main industrial revolution. The main industrial revolution begins with the steam engine. The first form of the steam engine was the crude and wasteful Newcomen engine, which was used for pumping mines. In the middle of the eighteenth century there were abortive attempts to use it for generating power, by making it pump water into elevated reservoirs, and employing the fall of this water to turn water wheels. Such clumsy devices became obsolete with the introduction of the perfected engines of Watt, which were employed quite early in their history for factory purposes as well as for mine pumping. The end of the eighteenth century saw the steam engine thoroughly established in industry, and the promise of the steamboat on the rivers and of steam traction on land was not far away.

The first place where steam power came into practical use was in replacing one of the most brutal forms of human or animal labor: pumping of water out of mines. At best, this had been done by draft animals, by crude machines turned by horses. At worst, as in the silver mines of New Spain, it was done by the labor of human slaves. It is a work that is never finished and which can never be interrupted without the possibility of closing down the mine forever. The use of the steam engine to replace this servitude must certainly be regarded as a great humanitarian step forward.

However, slaves do not only pump mines: they also drag loaded riverboats upstream. A second great triumph of the steam engine was the invention of the steamboat, and in particular of the river steamboat. The steam engine at sea was for many years but a supplement of questionable value to the sails carried by every seagoing steamboat; but it was steam transportation on the Mississippi which opened up the interior of the United States. Like the steamboat, the steam locomotive started where it is now dying, as a means of hauling heavy freight.

The next place where the industrial revolution made itself felt, perhaps a little later than in the field of the heavy labor of mine workers, and simultaneously with the revolution in transportation, was in the textile industry. This was already a sick industry. Even before the power spindle and the power looms, the condition of the spinners and the weavers left much to be desired. The bulk of production which they could perform fell far short of the demands of the day. It might thus appear to have been scarcely possible to conceive that the transition to the machine could have worsened their condition; but worsen it, it most certainly did.

The beginnings of textile-machine development go back of the steam engine. The stocking frame has existed in a form worked by hand ever since the time of Queen Elizabeth. Machine spinning first became necessary in order to furnish warps for hand looms. The complete mechanization of the textile industry, covering weaving as well as spinning, did not occur until the beginning of the nineteenth century. The first textile machines were for hand operation, although the use of horsepower and water power followed very quickly. Part of the impetus behind the development of the Watt engine, as contrasted with the Newcomen engine, was the desire to furnish power in the rotary form needed for textile purposes.

The textile mills furnished the model for almost the whole course of the mechanization of industry. On the social side, they began the transfer of the workers from the home to the factory and from the country to the city. There was an exploitation of the labor of children and women to an extent, and of a brutality scarcely conceivable at the present time—that is, if we forget the South African diamond mines and ignore the new industrialization of China and India and the general terms of plantation labor in almost every country. A great deal of this was due to the fact that new techniques had produced new responsibilities, at a time at which no code had yet arisen to take care of these responsibilities. There was, however, a phase which was of greater technical than moral significance. By this, I mean that a great many of the disastrous consequences and phases of the earlier part of the industrial revolution were not so much due to any moral obtuseness or iniquity on the part of those concerned, as to certain technical features which were inherent in the early means of industrialization, and which the later history of technical development has thrust more or less into the background. These technical determinants of the direction which the early industrial revolution took, lay in the very nature of early steam power and its transmission. The steam engine used fuel very uneconomically by modern standards, although this is not as important as it might seem, considering the fact that early engines had none of the more modern type with which to compete. However, among themselves they were much more economical to run on a large scale than on a small one. In contrast with the prime mover, the textile machine, whether it be loom or spindle, is a comparatively light machine, and uses little power. It was therefore economically necessary to assemble these machines in large factories, where many looms and spindles could be run from one steam engine.

At that time the only available means of transmission of power were mechanical. The first among these was the line of shafting, supplemented by the belt and the pulley. Even as late as the time of my own childhood, the typical picture of a factory was that of a great shed with long lines of shafts suspended from the rafters, and pulleys connected by belts to the individual machines. This sort of factory still exists; although in very many cases it has given way to the modern arrangement where the machines are driven individually by electric motors.

Indeed this second picture is the typical one at the present time. The trade of the millwright has taken on a totally new form. Here there is an important fact relevant to the whole history of invention. It was exactly these millwrights and other new craftsmen of the machine age who were to develop the inventions that are at the foundation of our patent system. Now, the mechanical connection of machines involves difficulties that are quite serious, and not easy to cover by any simple mathematical formulation. In the first place, long lines of shafting either have to be well aligned, or to employ ingenious modes of connection, such as universal joints or parallel couplings, which allow for a certain amount of freedom. In the second place, the long lines of bearings needed for such shafts are very high in their power consumption. In the individual machine, the rotating and reciprocating parts are subject to similar demands of rigidity, and to similar demands that the number of bearings must be reduced as much as possible for the sake of low power consumption and simple manufacture. These prescriptions are not easily filled on the basis of general formulas, and they offer an excellent opportunity for ingenuity and inventive skill of the old-fashioned artisan sort.

It is in view of this fact that the change-over in engineering between mechanical connections and electrical connections has had so great an effect. The electrical motor is a mode of distributing power which is very convenient to construct in small sizes, so that each machine may have its own motor. The transmission losses in the wiring of a factory are relatively low, and the efficiency of the motor itself is relatively high. The connection of the motor with its wiring is not necessarily rigid, nor does it consist of many parts. There are still motives of traffic and convenience which may induce us to continue the custom of mounting the different machines of an industrial process in a single factory; but the need of connecting all the machines to a single source of power is no longer a serious reason for geographical proximity. In other words, we are now in a position to return to cottage industry, in places where it would otherwise be suitable.

I do not wish to insist that the difficulties of mechanical transmission were the only cause of the shed factories and of the demoralization they produced. Indeed, the factory system started before the machine system, as a means of introducing discipline into the highly undisciplined home industry of the individual workers, and of keeping up standards of production.

It is true, however, that these non-mechanical factories were very soon superseded by mechanical ones, and that probably the worst social effects of urban crowding and of rural depopulation took place in the machine factory. Furthermore, if the fractional horsepower motor had been available from the start and could have increased the unit of production of a cottage worker, it is highly probable that a large part of the organization and discipline needed for successful large-scale production could have been superimposed on such home industries as spinning and weaving.

If it should be so desired, a single piece of machinery may now contain several motors, each introducing power at the proper place. This relieves the designer of much of the need for the ingenuity in mechanical design which he would otherwise have been compelled to use. In an electrical design, the mere problem of the connection of the parts seldom involves much difficulty that does not lend itself to easy mathematical formulation and solution. The inventor of linkages has been superseded by the computor of circuits. This is an example of the way in which the art of invention is conditioned by the existing means.

In the third quarter of the last century, when the electric motor was first used in industry, it was at first supposed to be nothing more than an alternative device for carrying out existing industrial techniques. It was probably not foreseen that its final effect would be to give rise to a new concept of the factory.

That other great electrical invention, the vacuum tube, has had a similar history. Before the invention of the vacuum tube, it took many separate mechanisms to regulate systems of great power. Indeed, most of the regulatory means themselves employed considerable power. There were exceptions to this, but only in specific fields, such as the steering of ships.

As late as 1915, I crossed the ocean on one of the old ships of the American Line. It belonged to the transitional period when ships still carried sails, as well as a pointed bow to carry a bowsprit. In a well-deck not far aft of the main superstructure, there was a formidable engine, consisting of four or five six-foot wheels with hand-spokes. These wheels were supposed to control the ship in the event that its automatic steering engine broke down. In a storm, it would have taken ten men or more, exerting their full strength, to keep that great ship on its course.

This was not the usual method of control of the ship, but an emergency replacement, or as sailors call it, a "jury steering wheel." For normal control, the ship carried a steering engine which translated the relatively small forces of the quartermaster at the wheel into the movement of the massive rudder. Thus even on a purely mechanical basis, some progress had been made toward the solution of the problem of amplification of forces or torques. Nevertheless, at that time, this solution of the amplification problem did not range over extreme differences between the levels of input and of output, nor was it embodied in a convenient universal type of apparatus.

The most flexible universal apparatus for amplifying small energy-levels into high energy-levels is the vacuum tube, or electron valve. The history of this is interesting, though it is too complex for us to discuss here. It is however amusing to reflect that the invention of the electron valve originated in Edison's greatest scientific discovery and perhaps the only one which he did not capitalize into an invention.

He observed that when an electrode was placed inside an electric lamp, and was taken as electrically positive with respect to the filament, then a current would flow, if the filament were heated, but not otherwise. Through a series of inventions by other people, this led to a more effective way than any known before of controlling a large current by a small voltage. This is the basis of the modern radio industry, but it is also an industrial tool which is spreading widely into new fields. It is thus no longer necessary to control a process at high energy-levels by a mechanism in which the important details of control are carried out at these levels. It is quite possible to form a certain pattern of behavior response at levels much lower even than those found in usual radio sets, and then to employ a series of amplifying tubes to control by this apparatus a machine as heavy as a steel-rolling mill. The work of discriminating and of forming the pattern of behavior for this is done under conditions in which the power losses are insignificant, and yet the final employment of this discriminatory process is at arbitrarily high levels of power.

It will be seen that this is an invention which alters the fundamental conditions of industry, quite as vitally as the transmission and subdivision of power through the use of the small electric motor. The study of the pattern of behavior is transferred to a special part of the instrument in which power-economy is of very little importance. We have thus deprived of much of their importance the dodges and devices previously used to insure that a mechanical linkage should consist of the fewest possible elements, as well as the devices used to minimize friction and lost motion. The design of machines involving such parts has been transferred from the domain of the skilled shopworker to that of the research-laboratory man; and in this he has all the available tools of circuit theory to replace a mechanical ingenuity of the old sort. Invention in the old sense has been supplanted by the intelligent employment of certain laws of nature. The step from the laws of nature to their employment has been reduced by a hundred times.

I have previously said that when an invention is made, a considerable period generally elapses before its full implications are understood. It was long before people became aware of the full impact of the airplane on international relations and on the conditions of human life. The effect of atomic energy on mankind and the future is yet to be assessed, although many observers insist that it is merely a new weapon like all older weapons.

The case of the vacuum tube was similar. In the beginning, it was regarded merely as an extra tool to supplement the already existing techniques of telephonic communication. The electrical engineers first mistook its real importance to such an extent that for years vacuum tubes were relegated simply to a particular part of the communication network. This part was connected with other parts consisting only of the traditional so-called inactive circuit elements—the resistances, the capacitances, and the inductances. Only since the war have engineers felt free enough in their employment of vacuum tubes to insert them where necessary, in the same way they had previously inserted passive elements of these three kinds.

The vacuum tube was first used to replace previously existing components of long-distance telephone circuits and wireless telegraphy. It was not long, however, before it became clear that the radio-telephone had achieved the stature of the radio-telegraph, and that broadcasting was possible. Let not the fact that this great triumph of invention has largely been given over to the soap-opera and the hillbilly singer, blind one to the excellent work that was done in developing it, and to the great civilizing possibilities which have been perverted into a national medicine-show.

Though the vacuum tube received its debut in the communications industry, the boundaries and extent of this industry were not fully understood for a long period. There were sporadic uses of the vacuum tube and of its sister invention, the photoelectric cell, for scanning the products of industry; as for example, for regulating the thickness of a web coming out of a paper machine, or for inspecting the color of a can of pineapples. These uses did not as yet form a reasoned new technique, nor were they associated in the engineering mind with the vacuum tubes other function, communication.

All this changed in the war. One of the few things gained from the great conflict was the rapid development of invention, under the stimulus of necessity and the unlimited employment of money; and above all, the new blood called in to industrial research. At the beginning of the war, our greatest need was to keep England from being knocked out by an overwhelming air attack. Accordingly, the anti-aircraft cannon was one of the first objects of our scientific war effort, especially when combined with the airplane-detecting device of radar or ultra-high-frequency Hertzian waves. The technique of radar used the same modalities as the existing technique of radio besides inventing new ones of its own. It was thus natural to consider radar as a branch of communication theory.

Besides finding airplanes by radar it was necessary to shoot them down. This involves the problem of fire control. The speed of the airplane has made it necessary to compute the elements of the trajectory of the anti-aircraft missile by machine, and to give the predicting machine itself communication functions which had previously been assigned to human beings. Thus the problem of anti-aircraft fire control made a new generation of engineers familiar with the notion of a communication addressed to a machine rather than to a person. In our chapter in language, we have already mentioned another field in which for a considerable time this notion had been familiar to a limited group of engineers: the field of the automatic hydroelectric power station.

During the period immediately preceding World War II other uses were found for the vacuum tube coupled directly to the machine rather than to the human agent. Among these were more general applications to computing machines. The concept of the large-scale computing machine as developed by Vannevar Bush among others was originally a purely mechanical one. The integration was done by rolling disks engaging one another in a frictional manner; and the interchange of outputs and inputs between these disks was the task of a classical train of shafts and gears.

The mother idea of these first computing machines is much older than the work of Vannevar Bush. In certain respects it goes back to the work of Babbage early in the last century. Babbage had a surprisingly modern idea for a computing machine, but his mechanical means fell far behind his ambitions. The first difficulty he met, and with which he could not cope, was that a long train of gears requires considerable power to run it, so that its output of power and torque very soon becomes too small to actuate the remaining parts of the apparatus. Bush saw this difficulty and overcame it in a very ingenious way. Besides the electrical amplifiers depending on vacuum tubes and on similar devices, there are certain mechanical torque-amplifiers which are familiar, for example, to everyone acquainted with ships and the unloading of cargo. The stevedore raises the cargo-slings by taking a purchase on his load around the drum of a donkey-engine or cargo-hoist. In this way, the tension which he exerts mechanically is increased by a factor which grows extremely rapidly with the angle of contact between his rope and the rotating drum. Thus one man is able to control the lifting of a load of many tons.

This device is fundamentally a force- or torque-amplifier. By an ingenious bit of design, Bush inserted such mechanical amplifiers between the stages of his computing machine, and thus was able to do effectively the sort of thing which had only been a dream for Babbage.

In the early days of Vannevar Bush's work, before there were any high speed automatic controls in factories, I had become interested in the problem of a partial differential equation. Bush's work had concerned the ordinary differential equation in which the independent variable was the time, and duplicated in its time course the course of the phenomena it was analyzing, although possibly at a different rate. In the partial differential equation, the quantities which take the place of the time are spread out in space, and I suggested to Bush that in view of the technique of television scanning which was then developing at high speed, we would, ourselves, have to consider such a technique to represent the many variables of, let us say, space, against the one variable of time. A computing machine so designed would have to work at extremely high speed, which to my mind put mechanical processes out of the question and threw us back on electronic processes. In such a machine, moreover, all data would have to be written, read, and erased with a speed compatible with that of the other operations of the machine; and in addition to including an arithmetical mechanism, it would need a logical mechanism as well and would have to be able to handle problems of programming on a purely logical and automatic basis. The notion of programming in the factory had already become familiar through the work of Taylor and the Gilbreths on time study, and was ready to be transferred to the machine. This offered considerable difficulty of detail, but no great difficulty of principle. I was thus convinced as far back as 1940 that the automatic factory was on the horizon, and I so informed Vannevar Bush. The consequent development of automatization, both before and after the publication of the first edition of this book, has convinced me that I was right in my judgment and that this development would be one of the great factors in conditioning the social and technical life of the age to come, the keynote of the second industrial revolution.

In one of its earlier phases the Bush Differential Analyzer performed all the principal amplification functions. It used electricity only to give power to the motors running the machine as a whole. This state of computing-mechanisms was intermediate and transitory. It very soon became clear that amplifiers of an electric nature, connected by wires rather than by shafts, were both less expensive and more flexible than mechanical amplifiers and connections. Accordingly, the later forms of Bush's machine made use of vacuum-tube devices. This has been continued in all their successors; whether they were what are now called analogy machines, which work primarily by the measurement of physical quantities, or digital machines, which work primarily by counting and arithmetic operations.

The development of these computing machines has been very rapid since the war. For a large range of computational work, they have shown themselves much faster and more accurate than the human computer. Their speed has long since reached such a level that any intermediate human intervention in their work is out of the question. Thus they offer the same need to replace human capacities by machine capacities as those which we found in the anti-aircraft computer. The parts of the machine must speak to one another through an appropriate language, without speaking to any person or listening to any person, except in the terminal and initial stages of the process. Here again we have an element which has contributed to the general acceptance of the extension to machines of the idea of communication.

In this conversation between the parts of a machine, it is often necessary to take cognizance of what the machine has already said. Here there enters the principle of feedback, which we have already discussed, and which is older than its exemplification in the ship's steering engine, and is at least as old, in fact, as the governor which regulates the speed of Watt's steam engine. This governor keeps the engine from running wild when its load is removed. If it starts to run wild, the balls of the governor fly upward from centrifugal action, and in their upward flight they move a lever which partly cuts off the admission of steam. Thus the tendency to speed up produces a partly compensatory tendency to slow down. This method of regulation received a thorough mathematical analysis at the hands of Clerk Maxwell in 1868.

Here feedback is used to regulate the velocity of a machine. In the ship's steering engine it regulates the position of the rudder. The man at the wheel operates a light transmission system, employing chains or hydraulic transmission, which moves a member in the room containing the steering engine. There is some sort of apparatus which notes the distance between this member and the tiller; and this distance controls the admission of steam to the ports of a steam steering-engine, or some similar electrical admission in the case of an electrical steering-engine. Whatever the particular connections may be, this change of admission is always in such a direction as to bring into coincidence the tiller and the member actuated from the wheel. Thus one man at the wheel can do with ease what a whole crew could do only with difficulty at the old manpower wheel.

We have so far given examples of where the feedback process takes primarily a mechanical form. However, a series of operations of the same structure can be carried out through electrical and even vacuum-tube means. These means promise to be the future standard method of designing control apparatus.

There has long been a tendency to render factories and machines automatic. Except for some special purpose, one would no longer think of producing screws by the use of the ordinary lathe, in which a mechanic must watch the progress of his cutter and regulate it by hand. The production of screws in quantity without serious human intervention is now the normal task of the ordinary screw machine. Although this does not make any special use of the process of feedback nor of the vacuum tube, it accomplishes a somewhat similar end. What the feedback and the vacuum tube have made possible is not the sporadic design of individual automatic mechanisms, but a general policy for the construction of automatic mechanisms of the most varied type. In this they have been reinforced by our new theoretical treatment of communication, which takes full cognizance of the possibilities of communication between machine and machine. It is this conjunction of circumstances which now renders possible the new automatic age.

The existing state of industrial techniques includes the whole of the results of the first industrial revolution, together with many inventions which we now see to be precursors of the second industrial revolution. What the precise boundary between these two revolutions may be, it is still too early to say. In its potential significance, the vacuum tube certainly belongs to an industrial revolution different from that of the age of power; and yet it is only at present that the true significance of the invention of the vacuum tube has been sufficiently realized to allow us to attribute the present age to a new and second industrial revolution.

Up to now we have been talking about the existing state of affairs. We have not covered more than a small part of the aspects of the previous industrial revolution. We have not mentioned the airplane, nor the bulldozer, together with the other mechanical tools of construction, nor the automobile, nor even one-tenth of those factors which have converted modern life to something totally unlike the life of any other period. It is fair to say, however, that except for a considerable number of isolated examples, the industrial revolution up to the present has displaced man and the beast as a source of power, without making any great impression on other human functions. The best that a pick-and-shovel worker can do to make a living at the present time is to act as a sort of gleaner after the bulldozer. In all important respects, the man who has nothing but his physical power to sell has nothing to sell which it is worth anyone's money to buy.

Let us now go on to a picture of a more completely automatic age. Let us consider what for example the automobile factory of the future will be like; and in particular the assembly line, which is the part of the automobile factory that employs the most labor. In the first place, the sequence of operations will be controlled by something like a modern high-speed computing machine. In this book and elsewhere, I have often said that the high-speed computing machine is primarily a logical machine, which confronts different propositions with one another and draws some of their consequences. It is possible to translate the whole of mathematics into the performance of a sequence of purely logical tasks. If this representation of mathematics is embodied in the machine, the machine will be a computing machine in the ordinary sense. However, such a computing machine, besides accomplishing ordinary mathematical tasks, will be able to undertake the logical task of channeling a series of orders concerning mathematical operations. Therefore, as present high-speed computing machines in fact do, it will contain at least one large assembly which is purely logical.

The instructions to such a machine, and here too I am speaking of present practice, are given by what we have called a taping. The orders given the machine may be fed into it by a taping which is completely predetermined. It is also possible that the actual contingencies met in the performance of the machine may be handed over as a basis of further regulation to a new control tape constructed by the machine itself, or to a modification of the old one. I have already explained how I think such processes are related to learning.

It may be thought that the present great expense of computing machines bars them from use in industrial processes; and furthermore that the delicacy of the work needed in their construction and the variability of their functions precludes the use of the methods of mass production in constructing them. Neither of these charges is correct. In the first place, the enormous computing machines now used for the highest level of mathematical work cost something of the order of hundreds of thousands of dollars. Even this price would not be forbidding for the control machine of a really large factory, but it is not the relevant price. The present computing machines are developing so rapidly that practically every one constructed is a new model. In other words, a large part of these apparently exorbitant prices goes into new work of design, and into new parts, which are produced by a very high quality of labor under the most expensive circumstances. If one of these computing machines were therefore established in price and model, and put to use in quantities of tens or twenties, it is very doubtful whether its price would be higher than tens of thousands of dollars. A similar machine of smaller capacity, not suited for the most difficult computational problems, but nevertheless quite adequate for factory control, would probably cost no more than a few thousand dollars in any sort of moderate-scale production.

Now let us consider the problem of the mass production of computing machines. If the only opportunity for mass production were the mass production of completed machines, it is quite clear that for a considerable period the best we could hope for would be a moderate-scale production. However, in each machine the parts are largely repetitive in very considerable numbers. This is true, whether we consider the memory apparatus, the logical apparatus, or the arithmetical subassembly. Thus production of a few dozen machines only, represents a potential mass production of the parts, and is accompanied by the same economic advantages.

It may still seem that the delicacy of the machines must mean that each job demands a special new model. This is also false. Given even a rough similarity in the type of mathematical and logical operations demanded of the mathematical and logical units of the machine, the over-all performance is regulated by the taping, or at any rate by the *original* taping. The taping of such a machine is a highly skilled task for a professional man of a very specialized type; but it is largely or entirely a once-for-all job, and need only be partly repeated when the machine is modified for a new industrial setup. Thus the cost of such a skilled technician will be distributed over a tremendous output, and will not really be a significant factor in the use of the machine.

The computing machine represents the center of the automatic factory, but it will never be the whole factory. On the one hand, it receives its detailed instructions from elements of the nature of sense organs, such as photoelectric cells, condensers for the reading of the thickness of a web of paper, thermometers, hydrogen-ion-concentration meters, and the general run of apparatus now built by instrument companies for the manual control of industrial processes. These instruments are already built to report electrically at remote stations. All they need to enable them to introduce their information into an automatic high-speed computer is a reading apparatus which will translate position or scale into a pattern of consecutive digits. Such apparatus already exists, and offers no great difficulty, either of principle or of constructional detail. The sense-organ problem is not new, and it is already effectively solved.

Besides these sense organs, the control system must

contain effectors, or components which act on the outer world. Some of these are of a type already familiar, such as valve-turning motors, electric clutches, and the like. Some of them will have to be invented, to duplicate more nearly the functions of the human hand as supplemented by the human eye. It is altogether possible in the machining of automobile frames to leave on certain metal lugs, machined into smooth surfaces as points of reference. The tool, whether it be a drill or riveter or whatever else we want, may be led to the approximate neighborhood of these surfaces by a photoelectric mechanism, actuated for example by spots of paint. The final positioning may bring the tool up against the reference surfaces, so as to establish a firm contact, but not a destructively firm one. This is only one way of doing the job. Any competent engineer can think of a dozen more.

Of course, we assume that the instruments which act as sense organs record not only the original state of the work, but also the result of all the previous processes. Thus the machine may carry out feedback operations, either those of the simple type now so thoroughly understood, or those involving more complicated processes of discrimination, regulated by the central control as a logical or mathematical system. In other words, the all-over system will correspond to the complete animal with sense organs, effectors, and proprioceptors, and not, as in the ultra-rapid computing machine, to an isolated brain, dependent for its experiences and for its effectiveness on our intervention.

The speed with which these new devices are likely to come into industrial use will vary greatly with the different industries. Automatic machines, which may not be precisely like those described here, but which perform roughly the same functions, have already come into extensive use in continuous-process industries like canneries, steel-rolling mills, and especially wire and tin-plate factories. They are also familiar in paper factories, which likewise produce a continuous output. Another place where they are indispensable is in that sort of factory which is too dangerous for any considerable number of workers to risk their lives in its control, and in which an emergency is likely to be so serious and costly that its possibilities should have been considered in advance, rather than left to the excited judgment of somebody on the spot. If a policy can be thought out in advance, it can be committed to a taping which will regulate the conduct to be followed in accordance with the readings of the instrument. In other words, such factories should be under a regime rather like that of the interlocking signals and switches of the railroad signal-tower. This regime is already followed in oil-cracking factories, in many other chemical works, and in the handling of the sort of dangerous materials found in the exploitation of atomic energy.

We have already mentioned the assembly line as a place for applying the same sorts of technique. In the assembly line, as in the chemical factory, or the continuous-process paper mill, it is necessary to exert a certain statistical control on the quality of the product. This control depends on a sampling process. These sampling processes have now been developed by Wald and others into a technique called *sequential analysis,* in which the sampling is no longer taken in a lump, but is a continuous process going along with the production. That which can be done then by a technique so standardized that it can be put in the hands of a statistical computer who does not understand the logic behind it, may also be executed by a computing machine. In other words, except again at the highest levels, the machine takes care of the routine statistical controls, as well as of the production process.

In general, factories have an accounting procedure which is independent of production, but insofar as the data which occur in cost-accounting come from the machine or assembly line, they may be fed directly into a computing machine. Other data may be fed in from time to time by human operators, but the bulk of the clerical work can be handled mechanically, leaving only the extraordinary details such as outside correspondence for human beings. But even a large part of the outside correspondence may be received from the correspondents on punched cards, or transferred to punched cards by extremely low-grade labor. From this stage on, everything may go by machine. This mechanization also may apply to a not inappreciable part of the library and filing facilities of an industrial plant.

In other words, the machine plays no favorites between manual labor and white-collar labor. Thus the possible fields into which the new industrial revolution is likely to penetrate are very extensive, and include all labor performing judgments of a low level, in much the same way as the displaced labor of the earlier industrial revolution included every aspect of human power. There will, of course, be trades into which the new industrial revolution will not penetrate either because the new control machines are not economical in industries on so small a scale as not to be able to carry the considerable capital costs involved, or because their work is so varied that a new taping will be necessary for almost every job. I cannot see automatic machinery of the judgment-replacing type coming into use in the corner grocery, or in the corner garage, although I can very well see it employed by the wholesale grocer and the automobile manufacturer. The farm laborer too, although he is beginning to be pressed by automatic machinery, is protected from the full pressure of it because of the ground he has to cover, the variability of the crops he must till, and the special conditions of weather and the like that he must meet. Even here, the large-scale or plantation farmer is becoming increasingly dependent on cotton-picking and weed-burning machinery, as the wheat farmer has long been dependent on the McCormick reaper. Where such machines may be used, some use of machinery of judgment is not inconceivable.

The introduction of the new devices and the dates at which they are to be expected are, of course, largely economic matters, on which I am not an expert. Short of any violent political changes or another great war, I should give a rough estimate that it will take the new tools ten to twenty years to come into their own. A war would change all this overnight. If we should engage in a war with a major power like Russia, which would make serious demands on the infantry, and consequently on our manpower, we may be hard put to keep up our industrial production. Under these circumstances, the matter of replacing human production by other modes may well be a life-or-death matter to the nation. We are already as far along in the process of developing a unified system of automatic control machines as we were in the development of radar in 1939. Just as the emergency of the Battle of Britain made it necessary to attack the radar problem in a massive manner, and to hurry up the natural development of the field by what may have been decades, so too, the needs of labor replacement are likely to act on us in a similar way in the case of another war. Personnel such as skilled radio amateurs, mathematicians, and physicists, who were so rapidly turned into competent electrical engineers for the purposes of radar design, is still available for the similar task of automatic-machine design. There is a new and skilled generation coming up, which they have trained.

Under these circumstances, the period of about two years which it took for radar to get onto the battlefield with a high degree of effectiveness is scarcely likely to be exceeded by the period of evolution of the automatic factory. At the end of such a war, the "know-how" needed to construct such factories will be common. There will even be a considerable backlog of equipment manufactured for the government, which is likely to be on sale or available to the industrialists. Thus a new war will almost inevitably see the automatic age in full swing within less than five years.

I have spoken of the actuality and the imminence of this new possibility. What can we expect of its economic and social consequences? In the first place, we can expect an abrupt and final cessation of the demand for the type of factory labor performing purely repetitive tasks. In the long run, the deadly uninteresting nature of the repetitive task may make this a good thing and the source of leisure necessary for man's full cultural development. It may also produce cultural results as trivial and wasteful as the greater part of those so far obtained from the radio and the movies.

Be that as it may, the intermediate period of the introduction of the new means, especially if it comes in the fulminating manner to be expected from a new war, will lead to an immediate transitional period of disastrous confusion. We have a good deal of experience as to how the industrialists regard a new industrial potential. Their whole propaganda is to the effect that it must not be considered as the business of the government but must be left open to whatever entrepreneurs wish to invest money in it. We also know that they have very few inhibitions when it comes to taking all the profit out of an industry that there is to be taken, and then letting the public pick up the pieces. This is the history of the lumber and mining industries, and is part of what we have called in another chapter the traditional American philosophy of progress.

Under these circumstances, industry will be flooded with the new tools to the extent that they appear to yield immediate profits, irrespective of what long-time damage they can do. We shall see a process parallel to the way in which the use of atomic energy for bombs has been allowed to compromise the very necessary potentialities of the long-time use of atomic power to replace our oil and coal supplies, which are within centuries, if not decades, of utter exhaustion. Note well that atomic bombs do not compete with power companies.

Let us remember that the automatic machine, whatever we think of any feelings it may have or may not have, is the precise economic equivalent of slave labor. Any labor which competes with slave labor must accept the economic conditions of slave labor. It is perfectly clear that this will produce an unemployment situation, in comparison with which the present recession and even the depression of the thirties will seem a pleasant joke. This depression will ruin many industries—possibly even the industries which have taken advantage of the new potentialities. However, there is nothing in the industrial tradition which forbids an industrialist to make a sure and quick profit, and to get out before the crash touches him personally.

Thus the new industrial revolution is a two-edged sword. It may be used for the benefit of humanity, but only if humanity survives long enough to enter a period in which such a benefit is possible. It may also be used to destroy humanity, and if it is not used intelligently it can go very far in that direction. There are, however, hopeful signs on the horizon. Since the publication of the first edition of this book, I have participated in two big meetings with representatives of business management, and I have been delighted to see that awareness on the part of a great many of those present of the social dangers of our new technology and the social obligations of those responsible for management to see that the new modalities are used for the benefit of man, for increasing his leisure and enriching his spiritual life, rather than merely for profits and the worship of the machine as a new brazen calf. There are many dangers still ahead, but the roots of good will are there, and I do not feel as thoroughly pessimistic as I did at the time of the publication of the first edition of this book.

X

SOME COMMUNICATION MACHINES
AND THEIR FUTURE

I devoted the last chapter to the problem of the industrial and social impact of certain control machines which are already beginning to show important possibilities for the replacement of human labor. However, there are a variety of problems concerning automata which have nothing whatever to do with our factory system but serve either to illustrate and throw light on the possibilities of communicative mechanisms in general, or for semi-medical purposes for the prosthesis and replacement of human functions which have been lost or weakened in certain unfortunate individuals. The first machine which we shall discuss was designed for theoretical purposes as an illustration to an earlier piece of work which had been done by me on paper some years ago, together with my colleagues, Dr. Arturo Rosenblueth and Dr. Julian Bigelow. In this work we conjectured that the mechanism of voluntary activity was of a feedback nature, and accordingly, we sought in the human voluntary activity for the characteristics of breakdown which feedback mechanisms exhibit when they are overloaded.

The simplest type of breakdown exhibits itself as an oscillation in a goal-seeking process which appears only when that process is actively invoked. This corresponds rather closely to the human phenomenon known as *intention tremor*, in which, for example, when the patient reaches for a glass of water, his hand swings wider and wider, and he cannot lift up the glass.

There is another type of human tremor which is in some ways diametrically opposite to intention tremor. It is known as *Parkinsonianism,* and is familiar to all of us as the shaking palsy of old men. Here the patient displays the tremor even at rest; and, in fact, if the disease is not too greatly marked, only at rest. When he attempts to accomplish a definite purpose this tremor subsides to such an extent that the victim of an early stage of Parkinsonianism can even be a successful eye surgeon.

The three of us associated this Parkinsonian tremor with an aspect of feedback slightly different from the feedback associated with the accomplishment of purpose. In order to accomplish a purpose successfully, the various joints which are not directly associated with purposive movement must be kept in such a condition of mild *tonus* or tension, that the final purposive contraction of the muscles is properly backed up. In order to do this, a secondary feedback mechanism is required, whose locus in the brain does not seem to be the cerebellum, which is the central control station of the mechanism which breaks down in intention tremor. This second sort of feedback is known as postural feedback.

It can be shown mathematically that in both cases of tremor the feedback is excessively large. Now, when we consider the feedback which is important in Parkinsonianism, it turns out that the voluntary feedback which regulates the main motion is in the opposite direction to the postural feedback as far as the motion of the parts regulated by the postural feedback is concerned. Therefore, the existence of a purpose tends to cut down the excessive amplification of postural feedback, and may very well bring it below the oscillation level. These things were very well known to us theoretically, but until recently we had not gone to the trouble of making a working model of them. However, it became desirable for us to construct a demonstration apparatus which would act according to our theories.

Accordingly, Professor J. B. Wiesner of the Electronics Laboratory of the Massachusetts Institute of Technology discussed with me the possibility of constructing a tropism machine or machine with a simple fixed built-in purpose, with parts sufficiently adjustable to show the main phenomena of voluntary feedback, and of what we have just called postural feedback, and their breakdown. At our suggestion, Mr. Henry Singleton took up the problem of building such a machine, and carried it to a brilliant and successful conclusion. This machine has two principal modes of action, in one of which it is positively photo-tropic and searches for light, and in the other of which it is negatively photo-tropic and runs away from light. We called the machine in its two respective functions, the *Moth* and the *Bedbug*. The machine consists of a little three-wheeled cart with a propelling motor on the rear axle. The front wheel is a caster steered by a tiller. The cart carries a pair of forwardly directed photo cells, one of which takes in the left quadrant, while the other takes in the right quadrant. These cells are the opposite arms of a bridge. The output of the bridge which is reversible, is put through an adjustable amplifier. After this it goes to a positioning motor which regulates the position of one contact with a potentiometer. The other contact is also regulated by a positioning motor which moves the tiller as well. The output of the potentiometer which represents the difference between the position of the two positioning motors leads through a second adjustable amplifier to the second positioning motor, thus regulating the tiller.

According to the direction of the output of the bridge, this instrument will be steered either toward the quadrant with more intense light or away from it. In either case, it automatically tends to balance itself. There is thus a feedback dependent on the source of light proceeding from the light to the photoelectric cells, and thence to the rudder control system, by which it finally regulates the direction of its own motion and changes the angle of incidence of the light.

This feedback tends to accomplish the purpose of either positive or negative photo-tropism. It is the analogue of a voluntary feedback, for in man we consider that a voluntary action is essentially a choice among tropisms. When this feedback is overloaded by increasing the amplification, the little cart or "the moth" or "the bedbug" according to the direction of its tropism will seek the light or avoid it in an oscillatory manner, in which the oscillations grow ever larger. This is a close analogue to the phenomenon of intention tremor, which is associated with injury to the cerebellum.

The positioning mechanism for the rudder contains a second feedback which may be considered as postural. This feedback runs from the potentiometer to the second motor and back to the potentiometer, and its zero point is regulated by the output of the first feedback. If this is overloaded, the rudder goes into a second sort of tremor. This second tremor appears in the absence of light: that is, when the machine is not given a purpose. Theoretically, this is due to the fact that as far as the second mechanism is concerned, the action of the first mechanism is antagonistic to its feedback, and tends to decrease its amount. This phenomenon in man is what we have described as Parkinsonianism.

I have recently received a letter from Dr. Grey Walter of the Burden Neurological Institute at Bristol, England, in which he expresses interest in "the moth" or "bedbug," and in which he tells me of a similar mechanism of his own, which differs from mine in having a determined but variable purpose. In his own language, "We have included features other than inverse feedback which gives to it an exploratory and ethical attitude to the universe as well as a purely tropistic one." The possibility of such a change in behavior pattern is discussed in the chapter of this book concerning learning, and this discussion is directly relevant to the Walter machine, although at present I do not know just what means he uses to secure such a type of behavior.

The moth and Dr. Walter's further development of a tropism machine seem to be at first sight exercises in virtuosity, or at most, mechanical commentaries to a philosophical text. Nevertheless, they have a certain definite usefulness. The United States Army Medical Corps has taken photographs of "the moth" to compare with photographs of actual cases of nervous tremor, so that they are thus of assistance in the instruction of army neurologists.

There is a second class of machines with which we have also been concerned which has a much more direct and immediately important medical value. These machines may be used to make up for the losses of the maimed and of the sensorily deficient, as well as to give new and potentially dangerous powers to the already powerful. The help of the machine may extend to the construction of better artificial limbs; to instruments to help the blind to read pages of ordinary text by translating the visual pattern into auditory terms; and to other similar aids to make them aware of approaching dangers and to give them freedom of locomotion. In particular, we may use the machine to aid the totally deaf. Aids of this last class are probably the easiest to construct; partly because the technique of the telephone is the best studied and most familiar technique of communication; partly because the deprivation of hearing is overwhelmingly a deprivation of one thing—free participation in human conversation; and partly because the useful information carried by speech can be compressed into such a narrow compass that it is not beyond the carrying-power of the sense of touch.

Some time ago, Professor Wiesner told me that he was interested in the possibility of constructing an aid for the totally deaf, and that he would like to hear my views on the subject. I gave my views, and it turned out that we were of much the same opinion. We were aware of the work that had already been done on visible speech at the Bell Telephone Laboratories, and its relation to their earlier work on the *Vocoder*. We knew that the *Vocoder* work gave us a measure of the amount of information which it was necessary to transmit for the intelligibility of speech more favorable than that of any previous method. We felt, however, that visible speech had two disadvantages; namely, that it did not seem to be easy to produce in a portable form, and that it made too heavy demands on the sense of vision, which is relatively more important for the deaf person than for the rest of us. A rough estimate showed that a transfer to the sense of touch of the principle used in the visible-speech instrument was possible, and this we decided should be the basis of our apparatus.

We found out very soon after starting that the investigators at the Bell Laboratories had also considered the possibility of a tactile reception of sound, and had included it in their patent application. They were kind enough to tell us that they had done no experimental work on it, and that they left us free to go ahead on our researches. Accordingly, we put the design and development of this apparatus into the hands of Mr. Leon Levine, a graduate student in the Electronics Laboratory. We foresaw that the problem of training would be a large part of the work necessary to bring our device into actual use, and here we had the benefit of the counsel of Dr. Alexander Bavelas of our Department of Psychology.

The problem of interpreting speech through another sense than that of hearing, such as the sense of touch, may be given the following interpretation from the point of view of language. As we have said, we may roughly distinguish three stages of language, and two intermediate translations, between the outside world and the subjective receipt of information. The first stage consists in the acoustic symbols taken physically as vibrations in the air; the second or phonetic stage consists in the various phenomena in the inner ear and the associated part of the nervous system; the third or semantic stage represents the transfer of these symbols into an experience of meaning.

In the case of the deaf person, the first and the third stages are still present, but the second stage is missing. However, it is perfectly conceivable that we may replace the second stage by one by-passing the sense of hearing and proceeding, for example, through the sense of touch. Here the translation between the first stage and the new second stage is performed, not by a physical-nervous apparatus which is born into us but by an artificial, humanly-constructed system. The translation between the new second stage and the third stage is not directly accessible to our inspection, but represents the formation of a new system of habits and responses, such as those we develop when we learn to drive a car. The present status of our apparatus is this: the transition between the first and the new second stage is well under control, although there are certain technical difficulties still to be overcome. We are making studies of the learning process; that is, of the transition between the second and third stages; and in our opinion, they seem extremely promising. The best result that we can show as yet, is that with a learned vocabulary of twelve simple words, a run of eighty random repetitions was made with only six errors.

In our work, we had to keep certain facts always in mind. First among these, as we have said, is the fact that hearing is not only a sense of communication, but a sense of communication which receives its chief use in establishing a *rapport* with other individuals. It is also a sense corresponding to certain communicative activities on our part: namely, those of speech. Other uses of hearing are important, such as the reception of the sounds of nature and the appreciation of music, but they are not so important that we should consider a person as socially deaf if he shared in the ordinary, interpersonal communication by speech, and in no other use of hearing. In other words, hearing has the property that if we are deprived of all its uses except that of speech-communication with other people, we should still be suffering under a minimal handicap.

For the purpose of sensory prosthesis, we must consider the entire speech process as a unit. How essential this is is immediately observed when we consider the speech of deaf-mutes. With most deaf-mutes, a training in lip-reading is neither impossible nor excessively difficult, to the extent that such persons can achieve a quite tolerable proficiency in receiving speech-messages from others. On the other hand, and with very few exceptions, and these the result of the best and most recent training, the vast majority of deaf-mutes, though they can learn how to use their lips and mouths to produce sound, do so with a grotesque and harsh intonation, which represents a highly inefficient form of sending a message.

The difficulties lie in the fact that for these people the act of conversation has been broken into two entirely separate parts. We may simulate the situation for the normal person very easily if we give him a telephone-communication-system with another person, in which his own speech is not transmitted by the telephone to his own ears. It is very easy to construct such dead-microphone transmission systems, and they have actually been considered by the telephone companies, only to be rejected because of the frightful sense of frustration they cause, especially the frustration of not knowing how much of one's own voice gets onto the line. People using a system of this sort are always forced to yell at the top of their voices, to be sure that they have missed no opportunity to get the message accepted by the line.

We now come back to ordinary speech. We see that the processes of speech and hearing in the normal person have never been separated; but that the very learning of speech has been conditioned by the fact that each individual hears himself speak. For the best results it is not enough that the individual hear himself speak at widely separated occasions, and that he fill in the gaps between these by memory. A good quality of speech can only be attained when it is subject to a continuous monitoring and self-criticism. Any aid to the totally deaf must take advantage of this fact, and although it may indeed appeal to another sense, such as that of touch, rather than to the missing sense of hearing, it must resemble the electric hearing-aids of the present day in being portable and continuously worn.

The further philosophy of prosthesis for hearing depends on the amount of information effectively used in hearing. The crudest evaluation of this amount involves the estimate of the maximum that can be communicated over a sound range of 10,000 cycles, and an amplitude range of some 80 decibels. This load of communication, however, while it marks the maximum of what the ear could conceivably do, is much too great to represent the effective information given by speech in practice. In the first place, speech of telephone quality does not involve the transmission of more than 3000 cycles; and the amplitude range is certainly not more than from 5 to 10 decibels; but even here, while we have not exaggerated what is transmitted to the ear, we are still grossly exaggerating what is used by the ear and brain to reconstitute recognizable speech.

We have said that the best work done on this problem of estimation is the *Vocoder* work of the Bell Telephone Laboratories. It may be used to show that if human speech is properly divided into not more than five bands, and if these are rectified so that only their form-envelopes or outer shapes are perceived, and are used to modulate quite arbitrary sounds within their frequency range, then if these sounds are finally added up, the original speech is recognizable as speech and almost recognizable as the speech of a particular individual. Nevertheless the amount of possible information transmitted, used or unused, has been cut to not more than a tenth or hundredth of the original potential information present.

When we distinguish between used and unused information in speech, we distinguish between the maximum coding capacity of speech as received by the ear, and the maximum capacity that penetrates through the cascade network of successive stages consisting of the ear followed by the brain. The first is only relevant to the transmission of speech through the air and through intermediate instruments like the telephone, followed by the ear itself, but not by whatever apparatus in the brain is used in the understanding of speech. The second refers to the transmitting power of the entire complex—air—telephone—ear—brain. Of course, there may be finer shades of inflection which do not get through to the over-all narrow-band transmission system of which we are speaking, and it is hard to evaluate the amount of lost information carried by these; but it seems to be relatively small. This is the idea behind the *Vocoder*. The earlier engineering estimates of information were defective in that they ignored the terminating element of the chain from the air to the brain.

In appealing to the other senses of a deaf person, we must realize that apart from sight, all others are inferior to it, and transmit less information per unit time. The only way we can make an inferior sense like touch work with maximum efficiency is to send through it not the full information that we get through hearing, but an edited portion of that hearing suitable for the understanding of speech. In other words, we replace part of the function that the cortex normally performs after the reception of sound, by a filtering of our information before it goes through the tactile receptors. We thus transfer part of the function of the cortex of the brain to an artificial external cortex. The precise way we do this in the apparatus we are considering is by separating the frequency bands of speech as in the *Vocoder*, and then by transmitting these different rectified bands to spatially distant tactile regions, after they have been used to modulate vibrations of frequencies easily perceived by the skin. For example, five bands may be sent respectively to the thumb and four fingers of one hand.

This gives us the main ideas of the apparatus needed for the reception of intelligible speech through sound vibrations transformed electrically into touch. We have gone far enough already to know that the patterns of a considerable number of words are sufficiently distinct from one another, and sufficiently consistent among a number of speakers, to be recognized without any great amount of speech training. From this point on, the chief direction of investigation must be that of the more thorough training of deaf-mutes in the recognition and the reproduction of sounds. On the engineering end, we shall have considerable problems concerning the portability of the apparatus, and the reduction of its energy demands, without any substantial loss of performance. These matters are all still *sub judice*. I do not wish to establish false and in particular premature hopes on the part of the afflicted and their friends, but I think it is safe to say that the prospect of success is far from hopeless.

Since the publication of the first edition of this book, new special devices for elucidating points in the theory of communication have been developed by other workers. I have already mentioned in an earlier chapter the homeostats of Dr. Ashby and the somewhat similar machines of Dr. Grey Walter. Here let me mention some earlier machines of Dr. Walter, somewhat similar to my "moth" or "bug," but which were built for a different purpose. For these phototopic machines, each element carries a light so that it can stimulate the others. Thus a number of them put into operation at the same time show certain groupings and mutual reactions which would be interpreted by most animal psychologists as social behavior if they were found encased in flesh and blood instead of brass and steel. It is the beginning of a new science of mechanical behavior even though almost all of it lies in the future.

Here at M.I.T. circumstances have made it difficult to carry work on the hearing glove much further during the last two years, although the possibility of its development still exists. Meanwhile, the theory—although not the detail of the device—has led to improvements in apparatus to allow the blind to get themselves through a maze of streets and buildings. This research is largely the work of Dr. Clifford M. Witcher, himself congenitally blind, who is an outstanding authority and technician in optics, electrical engineering, and the other fields necessary to this work.

A prosthetic device which looks hopeful but has not yet been subjected to any real development or final criticism is an artificial lung in which the activation of the breathing motor will depend on signals, electrical or mechanical, from the weakened but not destroyed breathing muscles of the patient. In this case, the normal feedback in the medulla and brain stem of the healthy person will be used even in the paralytic to supply the control of his breathing. Thus it is hoped that the so-called iron lung may no longer be a prison in which the patient forgets how to breathe, but will be an exerciser for keeping his residual faculties of breathing active, and even possibly of building them up to a point where he can breathe for himself and emerge from the machinery enclosing him.

Up to the present, we have been discussing machines which as far as the general public is concerned seem either to share the characteristic detachment from immediate human concerns of theoretical science or to be definitely beneficent aids to the maimed. We now come to another class of machines which possess some very sinister possibilities. Curiously enough, this class contains the automatic chess-playing machine.

Sometime ago, I suggested a way in which one might use the modern computing machine to play at least a passable game of chess. In this work, I am following up a line of thought which has a considerable history behind it. Poe discussed a fraudulent chess-playing machine due to Maelzel, and exposed it; showing that it was worked by a legless cripple inside. However, the machine I have in mind is a genuine one, and takes advantage of recent progress in computing machines. It is easy to make a machine that will play merely legal chess of a very poor brand; it is hopeless to try to make a machine to play perfect chess for such a machine would require too many combinations. Professor John von Neumann of the Institute for Advanced Studies at Princeton has commented on this difficulty. However, it is neither easy nor hopeless to make a machine which we can guarantee to do the best that can be done for a limited number of moves ahead, say two; and which will then leave itself in the position that is the most favorable in accordance with some more or less easy method of evaluation.

The present ultra-rapid computing machines may be set up to act as chess-playing machines, though a better machine might be made at an exorbitant price if we chose to put the work into it. The speed of these modern computing machines is enough so that they can evaluate every possibility for two moves ahead in the legal playing-time of a single move. The number of combinations increases roughly in geometrical progression. Thus the difference between playing out all possibilities for two moves and for three moves is enormous. To play out a game—something like fifty moves—is hopeless in any reasonable time. Yet for beings living long enough, as von Neumann has shown, it would be possible; and a game played perfectly on each side would lead, as a foregone conclusion, either always to a win for White, or always to a win for Black, or most probably always to a draw.

Mr. Claude Shannon of the Bell Telephone Laboratories has suggested a machine along the same lines as the two-move machine I had contemplated, but considerably improved. To begin with, his evaluation of the final position after two moves would make allowances for the control of the board, for the mutual protection of the pieces, etc., as well as the number of pieces, check, and checkmate. Then too, if at the end of two moves, the game should be unstable, by the existence of check, or of an important piece in a position to be taken, or of a fork, the mechanical player would automatically play a move or two ahead until stability should be reached. How much this would slow the game, lengthening each move beyond the legal limit, I do not know; although I am not convinced that we can go very far in this direction without getting into time trouble at our present speeds.

I am willing to accept Shannon's conjecture that such a machine would play chess of a high amateur level and even possibly of a master level. Its game would be stiff and rather uninteresting, but much safer than that of any human player. As Shannon points out, it is possible to put enough chance in its operation to prevent its constant defeat in a.purely systematic way by a given rigid sequence of plays. This chance or uncertainty may be built into the evaluation of terminal positions after two moves.

The machine would play gambits and possibly end games like a human player from the store of standard gambits and end games. A better machine would store on a tape every game it had ever played and would supplement the processes which we have already indicated by a search through all past games to find something apropos: in short, by the power of learning. Though we have seen that machines can be built to learn, the technique of building and employing these machines is still very imperfect. The time is not yet ripe for the design of a chess-playing machine on learning principles, although it probably does not lie very far in the future.

A chess-playing machine which learns might show a great range of performance, dependent on the quality of the players against whom it had been pitted. The best way to make a master machine would probably be to pit it against a wide variety of good chess players. On the other hand, a well-contrived machine might be more or less ruined by the injudicious choice of its opponents. A horse is also ruined if the wrong riders are allowed to spoil it.

In the learning machine, it is well to distinguish what the machine can learn and what it cannot. A machine may be built either with a statistical preference for a certain sort of behavior, which nevertheless admits the possibility of other behavior; or else certain features of its behavior may be rigidly and unalterably determined. We shall call the first sort of determination *preference*, and the second sort of determination *constraint*. For example, if the rules of legal chess are not built into a chess-playing machine as constraints, and if the machine is given the power to learn, it may change without notice from a chess-playing machine into a machine doing a totally different task. On the other hand, a chess-playing machine with the rules built in as constraints may still be a learning machine as to tactics and policies.

The reader may wonder why we are interested in chess-playing machines at all. Are they not merely another harmless little vanity by which experts in design seek to show off their proficiency to a world which they hope will gasp and wonder at their accomplishments? As an honest man, I cannot deny that a certain element of ostentatious narcissism is present in me, at least. However, as you will soon see, it is not the only element active here, nor is it that which is of the greatest importance to the non-professional reader.

Mr. Shannon has presented some reasons why his researches may be of more importance than the mere design of a curiosity, interesting only to those who are playing a game. Among these possibilities, he suggests that such a machine may be the first step in the construction of a machine to evaluate military situations and to determine the best move at any specific stage. Let no man think that he is talking lightly. The great book of von Neumann and Morgenstern on the *Theory of Games* has made a profound impression on the world, and not least in Washington. When Mr. Shannon speaks of the development of military tactics, he is not talking moonshine, but is discussing a most imminent and dangerous contingency.

In the well-known Paris journal, *Le Monde*, for December 28, 1948, a certain Dominican friar, Père Dubarle, has written a very penetrating review of my book *Cybernetics*. I shall quote a suggestion of his which carries out some of the dire implications of the chess-playing machine grown up and encased in a suit of armor.

> One of the most fascinating prospects thus opened is that of the rational conduct of human affairs, and in particular of those which interest communities and seem to present a certain statistical regularity, such as the human phenomena of the development of opinion. Can't one imagine a machine to collect this or that type of information, as for example information on production and the market; and then to determine as a function of the average psychology of human beings, and of the quantities which it is possible to measure in a determined instance, what the most probable development of the situation might be? Can't one even conceive a State apparatus covering all systems of political decisions, either under a regime of many states distributed over the earth,

or under the apparently much more simple regime of a human government of this planet? At present nothing prevents our thinking of this. We may dream of the time when the *machine à gouverner* may come to supply—whether for good or evil—the present obvious inadequacy of the brain when the latter is concerned with the customary machinery of politics.

At all events, human realities do not admit a sharp and certain determination, as numerical data of computation do. They only admit the determination of their probable values. A machine to treat these processes, and the problems which they put, must therefore undertake the sort of probabilistic, rather than deterministic thought, such as is exhibited for example in modern computing machines. This makes its task more complicated, but does not render it impossible. The prediction machine which determines the efficacy of anti-aircraft fire is an example of this. Theoretically, time prediction is not impossible; neither is the determination of the most favorable decision, at least within certain limits. The possibility of playing machines such as the chess-playing machine is considered to establish this. For the human processes which constitute the object of government may be assimilated to games in the sense in which von Neumann has studied them mathematically. Even though these games have an incomplete set of rules, there are other games with a very large number of players, where the data are extremely complex. The *machines à gouverner* will define the State as the best-informed player at each particular level; and the State is the only supreme co-ordinator of all partial decisions. These are enormous privileges; if they are acquired scientifically, they will permit the State under all circumstances to beat every player of a human game other than itself by offering this dilemma: either immediate ruin, or planned co-operation. This will be the consequences of the game itself without outside violence. The lovers of the best of worlds have something indeed to dream of!

Despite all this, and perhaps fortunately, the *machine à gouverner* is not ready for a very near tomorrow. For outside of the very serious problems which the volume of information to be collected and to be treated rapidly still put, the problems of the stability of prediction remain beyond what we can seriously dream of controlling. For human processes are assimilable to games with incompletely defined rules, and above all, with the rules themselves functions of the time. The variation of the rules depends both on the effective detail of the situations engendered by the game itself, and on the system of psychological reactions of the players in the face of the results obtained at each instant.

It may even be more rapid than these. A very good example of this seems to be given by what happened to the Gallup Poll in the 1948 election. All this not only tends to complicate the degree of the factors which influence prediction, but perhaps to make radically sterile the mechanical manipulation of human situations. As far as one can judge, only two conditions here can guarantee stabilization in the mathematical sense of the term. These are, on the one hand, a sufficient ignorance on the part of the mass of the players exploited by a skilled player, who moreover may plan a method of paralyzing the consciousness of the masses; or on the other, sufficient good-will to allow one, for the sake of the stability of the game, to refer his decisions to one or a few players of the game who have arbitrary privileges. This is a hard lesson of cold mathematics, but it throws a certain light on the adventure of our century: hesitation between an indefinite turbulence of human affairs and the rise of a prodigious Leviathan. In comparison with this, Hobbes' *Leviathan* was nothing but a pleasant joke. We are running the risk nowadays of a great World State, where deliberate and conscious primitive injustice may be the only possible condition for the statistical happiness of the masses: a world worse than hell for every clear mind. Perhaps it would not be a bad idea for the teams at present creating cybernetics to add to their *cadre* of technicians, who have come from all horizons of science, some serious anthropologists, and perhaps a philosopher who has some curiosity as to world matters.

The *machine à gouverner* of Père Dubarle is not frightening because of any danger that it may achieve autonomous control over humanity. It is far too crude and imperfect to exhibit a one-thousandth part of the purposive independent behavior of the human being. Its real danger, however, is the quite different one that such machines, though helpless by themselves, may be used by a human being or a block of human beings to increase their control over the rest of the human race or that political leaders may attempt to control their populations by means not of machines themselves but through political techniques as narrow and indifferent to human possibility as if they had, in fact, been conceived mechanically. The great weakness of the machine—the weakness that saves us so far from being dominated by it—is that it cannot yet take into account the vast range of probability that characterizes the human situation. The dominance of the machine presupposes a society in the last stages of increasing entropy, where probability is negligible and where the statistical differences among individuals are nil. Fortunately we have not yet reached such a state.

But even without the state machine of Père Dubarle we are already developing new concepts of war, of economic conflict, and of propaganda on the basis of von Neumann's *Theory of Games*, which is itself a communicational theory, as the developments of the 1950s have already shown. This theory of games, as I have said in an earlier chapter, contributes to the theory of language, but there are in existence government agencies bent on applying it to military and quasi-military aggressive and defensive purposes.

The theory of games is, in its essence, based on an arrangement of players or coalitions of players each of whom is bent on developing a strategy for accomplishing its purposes, assuming that its antagonists, as well as itself, are each engaging in the best policy for victory. This great game is already being carried on mechanistically, and on a colossal scale. While the philosophy behind it is probably not acceptable to our present opponents, the Communists, there are strong

signs that its possibilities are already being studied in
Russia as well as here, and that the Russians, not content with accepting the theory as we have presented it,
have conceivably refined it in certain important respects. In particular, much of the work, although not
all, which we have done on the theory of games, is
based on the assumption that both we and our opponents have unlimited capabilities and that the only
restrictions within which we play depend on what we
may call the cards dealt to us or the visible positions
on the chess board. There is a considerable amount
of evidence, rather in deed than in words, that the
Russians have supplemented this attitude to the world
game by considering the psychological limits of the
players and especially their fatigability as part of the
game itself. A sort of *machine à gouverner* is thus now
essentially in operation on both sides of the world conflict, although it does not consist in either case of a
single machine which makes policy, but rather of a
mechanistic technique which is adapted to the exigen-
cies of a machine-like group of men devoted to the
formation of policy.

Père Dubarle has called the attention of the scientist to the growing military and political mechanization of the world as a great superhuman apparatus
working on cybernetic principles. In order to avoid the
manifold dangers of this, both external and internal,
he is quite right in his emphasis on the need for the
anthropologist and the philosopher. In other words, we
must know as scientists what man's nature is and what
his built-in purposes are, even when we must wield
this knowledge as soldiers and as statesmen; and we
must know why we wish to control him.

When I say that the machine's danger to society is
not from the machine itself but from what man makes
of it, I am really underlining the warning of Samuel
Butler. In *Erewhon* he conceives machines otherwise
unable to act, as conquering mankind by the use of

men as the subordinate organs. Nevertheless, we must not take Butler's foresight too seriously, as in fact at his time neither he nor anyone around him could understand the true nature of the behavior of automata, and his statements are rather incisive figures of speech than scientific remarks.

Our papers have been making a great deal of American "know-how" ever since we had the misfortune to discover the atomic bomb. There is one quality more important than "know-how" and we cannot accuse the United States of any undue amount of it. This is "know-what" by which we determine not only how to accomplish our purposes, but what our purposes are to be. I can distinguish between the two by an example. Some years ago, a prominent American engineer bought an expensive player-piano. It became clear after a week or two that this purchase did not correspond to any particular interest in the music played by the piano but rather to an overwhelming interest in the piano mechanism. For this gentleman, the player-piano was not a means of producing music, but a means of giving some inventor the chance of showing how skillful he was at overcoming certain difficulties in the production of music. This is an estimable attitude in a second-year high-school student. How estimable it is in one of those on whom the whole cultural future of the country depends, I leave to the reader.

In the myths and fairy tales that we read as children we learned a few of the simpler and more obvious truths of life, such as that when a djinnee is found in a bottle, it had better be left there; that the fisherman who craves a boon from heaven too many times on behalf of his wife will end up exactly where he started; that if you are given three wishes, you must be very careful what you wish for. These simple and obvious truths represent the childish equivalent of the tragic view of life which the Greeks and many modern

Europeans possess, and which is somehow missing in this land of plenty.

The Greeks regarded the act of discovering fire with very split emotions. On the one hand, fire was for them as for us a great benefit to all humanity. On the other, the carrying down of fire from heaven to earth was a defiance of the Gods of Olympus, and could not but be punished by them as a piece of insolence towards their prerogatives. Thus we see the great figure of Prometheus, the fire-bearer, the prototype of the scientist; a hero but a hero damned, chained on the Caucasus with vultures gnawing at his liver. We read the ringing lines of Aeschylus in which the bound god calls on the whole world under the sun to bear witness to what torments he suffers at the hands of the gods.

The sense of tragedy is that the world is not a pleasant little nest made for our protection, but a vast and largely hostile environment, in which we can achieve great things only by defying the gods; and that this defiance inevitably brings its own punishment. It is a dangerous world, in which there is no security, save the somewhat negative one of humility and restrained ambitions. It is a world in which there is a condign punishment, not only for him who sins in conscious arrogance, but for him whose sole crime is ignorance of the gods and the world around him.

If a man with this tragic sense approaches, not fire, but another manifestation of original power, like the splitting of the atom, he will do so with fear and trembling. He will not leap in where angels fear to tread, unless he is prepared to accept the punishment of the fallen angels. Neither will he calmly transfer to the machine made in his own image the responsibility for his choice of good and evil, without continuing to accept a full responsibility for that choice.

I have said that the modern man, and especially the modern American, however much "know-how" he may have, has very little "know-what." He will accept the

superior dexterity of the machine-made decisions with
out too much inquiry as to the motives and principles
behind these. In doing so, he will put himself sooner
or later in the position of the father in W. W. Jacobs'
The Monkey's Paw, who has wished for a hundred
pounds, only to find at his door the agent of the company for which his son works, tendering him one
hundred pounds as a consolation for his son's death at
the factory. Or again, he may do it in the way of the
Arab fisherman in the *One Thousand and One Nights*,
when he broke the Seal of Solomon on the lid of the
bottle which contained the angry djinnee.

Let us remember that there are game-playing machines both of the Monkey's Paw type and of the type
of the Bottled Djinnee. Any machine constructed for
the purpose of making decisions, if it does not possess
the power of learning, will be completely literal-
minded. Woe to us if we let it decide our conduct,
unless we have previously examined the laws of its
action, and know fully that its conduct will be carried
out on principles acceptable to us! On the other hand,
the machine like the djinnee, which can learn and can
make decisions on the basis of its learning, will in no
way be obliged to make such decisions as we should
have made, or will be acceptable to us. For the man
who is not aware of this, to throw the problem of his
responsibility on the machine, whether it can learn or
not, is to cast his responsibility to the winds, and to
find it coming back seated on the whirlwind.

I have spoken of machines, but not only of machines
having brains of brass and thews of iron. When human
atoms are knit into an organization in which they are
used, not in their full right as responsible human beings, but as cogs and levers and rods, it matters little
that their raw material is flesh and blood. *What is used
as an element in a machine, is in fact an element in
the machine.* Whether we entrust our decisions to machines of metal, or to those machines of flesh and blood

which are bureaus and vast laboratories and armies and corporations, we shall never receive the right answers to our questions unless we ask the right questions. The *Monkey's Paw* of skin and bone is quite as deadly as anything cast out of steel and iron. The djinnee which is a unifying figure of speech for a whole corporation is just as fearsome as if it were a glorified conjuring trick.

The hour is very late, and the choice of good and evil knocks at our door.

XI

LANGUAGE, CONFUSION, AND JAM

In Chapter IV, I have referred to some very interesting work recently carried out by Dr. Benoit Mandelbrot of Paris and Professor Jacobson of Harvard on various phenomena of language including, among other things, a discussion of the optimum distribution of the length of words. It is not my intention to go into the detail of this work in the present chapter, but rather to develop the consequences of certain philosophical assumptions made by these two writers.

They consider communication to be a game played in partnership by the speaker and the listener against the forces of confusion, represented by the ordinary difficulties of communication and by some supposed individuals attempting to jam the communication. Literally speaking, the game theory of von Neumann, which is involved in this connection, concerns one team which is deliberately trying to get the message across, and another team which will resort to any strategy to jam the message. In the strict von Neumann theory of games this means that the speaker and listener cooperate on policy in view of the assumption that the jamming agency is adopting the best policy to confuse them, again under the assumption that the speaker and the listener have been using the best policy up to the present, and so on.

In more usual language, both the team of communicants and the jamming forces are at liberty to use the technique of bluffing to confound one another, and in general this technique will be used to prevent the other side from being able to act on a firm knowledge of the technique of one side. Both sides will then bluff, the jamming force in order to adapt themselves to new communication techniques developed by the communicating forces, and the communicating forces to outwit any policy already developed by the jamming forces. In this connection concerning scientific method Albert Einstein's remark that I quoted earlier is of the greatest significance. *"Der Herr Gott ist raffiniert, aber boshaft ist Er nicht."* "God may be subtle, but he isn't plain mean."

Far from being a cliché, this is a very profound statement concerning the problems of the scientist. To discover the secrets of nature requires a powerful and elaborate technique, but at least we can expect one thing—that as far as inanimate nature goes, any step forward that we may take will not be countered by a change of policy by nature for the deliberate purpose of confusing and frustrating us. There may indeed be certain limitations to this statement as far as living nature is concerned, for the manifestations of hysteria are often made in view of an audience, and with the intention, which is frequently unconscious, of bamboozling that audience. On the other hand, just as we seem to have conquered a germ disease, the germ may mutate and show traits which at least appear to have been developed with the deliberate intention of sending us back to the point where we have started.

These infractuousities of nature, no matter how much they may annoy the practitioner of the life sciences, are fortunately not among the difficulties to be contemplated by the physicist. Nature plays fair and if, after climbing one range of mountains, the physicist sees another on the horizon before him, it has not been deliberately put there to frustrate the effort he has already made.

It may seem superficially that even in the absence of a conscious or purposeful interference by nature, the policy of the research scientist should be to play it safe, and always act so that even a malicious and deceitful nature would not prevent his optimum acquisition and transfer of information. This point of view is unjustified. Communication in general, and scientific research in particular, involve a great deal of effort even if it is useful effort, and the fighting of bogies which are not there wastes effort which ought to be economized. We can not go through our communicative or scientific lives shadow-boxing with ghosts. Experience has pretty well convinced the working physicist that any idea of a nature which is not only difficult to interpret but which actively resists interpretation has not been justified as far as his past work is concerned, and therefore, to be an effective scientist, he must be naïve, and even deliberately naïve, in making the assumption that he is dealing with an honest God, and must ask his questions of the world as an honest man.

Thus the naïveté of a scientist, while it is a professional adaptation, is not a professional defect. A man who approaches science with the point of view of an officer of detective police would spend most of his time frustrating tricks that are never going to be played on him, trailing suspects who would be perfectly willing to give an answer to a direct question, and in general playing the fashionable cops-and-robbers game as it is now played within the realm of official and military science. I have not the slightest doubt that the present detective-mindedness of the lords of scientific administration is one of the chief reasons for the barrenness of so much present scientific work.

It follows almost by a syllogism that there are other professions besides that of the detective which can and do disqualify a man for the most effective scientific work, both by making him suspect nature of disingenuousness, and by making him disingenuous in his attitude to nature and to questions about nature. The soldier is trained to regard life as a conflict between man and man, but even he is not as tightly bound to this point of view as the member of a militant religious order—the soldier of the Cross, or of the Hammer and Sickle. Here the existence of a fundamentally propagandist point of view is much more important than the particular nature of the propaganda. It matters little whether the military band to which one has pledged oneself be that of Ignatius Loyola or of Lenin, so long as he considers it more important that his beliefs should be on the right side than that he should maintain his freedom and even his professional naïveté. He is unfitted for the highest flights of science no matter what his allegiance, as long as that allegiance is absolute. In this present day when almost every ruling force, whether on the right or on the left, asks the scientist for conformity rather than openness of mind, it is easy to understand how science has already suffered, and what further debasements and frustrations of science are to be expected in the future.

I have already pointed out that the devil whom the scientist is fighting is the devil of confusion, not of willful malice. The view that nature reveals an entropic tendency is Augustinian, not Manichaean. Its inability to undertake an aggressive policy, deliberately to defeat the scientist, means that its evil doing is the result of a weakness in his nature rather than of a specifically evil power that it may have, equal or inferior to the principles of order in the universe which, local and temporary as they may be, still are probably not too unlike what the religious man means by God. In Augustinianism, the black of the world is negative and is the mere absence of white, while in Manichaeanism, white and black belong to two opposed armies drawn up in line facing one another. There is a subtle emotional Manichaeanism implicit in all crusades, all jihads, and all wars of communism against the devil of capitalism.

The Augustinian position has always been difficult to maintain. It tends under the slightest perturbation to break down into a covert Manichaeanism. The emotional difficulty of Augustinianism shows itself in Milton's dilemma in *Paradise Lost:* If the devil is merely the creature of God and belongs to a world in which God is omnipotent, serving to point out some of the dark, confusing corners of the world, the great battle between the fallen angels and the forces of the Lord becomes about as interesting as a professional wrestling match. If Milton's poem is to have the dignity of being more than one of these groan-and-grunt exhibitions, the devil must be given a chance of winning, at least in his own estimation, even though it be no more than an outside chance. The devil's own words in *Paradise Lost* convey his awareness of the omnipotence of God and the hopelessness of fighting him, yet his actions indicate that at least emotionally he considers this fight a desperate, but not utterly useless, assertion of the rights of his hosts and of himself. Even the Augustinian devil must watch his step or he will be converted to Manichaeanism.

Any religious order which is based on the military model is under this same temptation to lapse into the Manichaean heresy. It has adopted as a simile for the forces which it is fighting that of an independent army which it is determined to defeat; but which could, at least conceivably, win the war and itself become the ruling force. For this reason, such an order or organization is intrinsically unsuited to encourage an Augustinian attitude within the scientist; and furthermore, it does not tend to rate limpid intellectual honesty very high in its scale of virtues. Against an insidious enemy who himself plays tricks, military stratagems are permissible. Thus a religious military order is almost bound to set a great value on obedience, confessions of faith, and all the restricting influences which hamstring the scientist.

It is true that nobody can speak for the Church but the Church itself, but it is equally true that those outside the Church may, and even must, have their own attitudes toward the organization and its claims. It is equally true that communism as an intellectual force is fundamentally what the Communists say it is, but their statements have a binding claim on us only as matters of the definition of an ideal and not as a description that we can act on of a specific organization or movement.

It appears that Marx's own view was Augustinian, and that evil was for him rather a lack of perfection than an autonomous positioned force fighting against good. Nevertheless, communism has grown up in an atmosphere of combat and conflict, and the general tendency seems to be to relegate the final Hegelian synthesis to which the Augustinian attitude toward evil is appropriate, to a future which, if not infinitely remote, has at least a very attenuated reference to what is happening at present.

Thus, for the present, and as a matter of practical conduct, both the camp of communism and many elements in the camp of the Church take attitudes which are definitely Manichaean. I have implied that Manichaeanism is a bad atmosphere for science. Curious as it may seem, this is because it is a bad atmosphere for faith. When we do not know whether a particular phenomenon we observe is the work of God or the work of Satan, the very roots of our faith are shaken. It is only under such a condition that it is possible to make a significant, willful choice between God and Satan, and this choice may lead to diabolism, or (in other words) to witchcraft. Furthermore, it is only in an atmosphere in which witchcraft is genuinely possible that witch-hunting flourishes as a significant activity. Thus it is no accident that Russia has had its Berias and that we have our McCarthys.

I have said that science is impossible without faith. By this I do not mean that the faith on which science depends is religious in nature or involves the acceptance of any of the dogmas of the ordinary religious creeds, yet without faith that nature is subject to law there can be no science. No amount of demonstration can ever prove that nature is subject to law. For all we know, the world from the next moment on might be something like the croquet game in *Alice in Wonderland*, where the balls are hedgehogs which walk off, the hoops are soldiers who march to other parts of the field, and the rules of the game are made from instant to instant by the arbitrary decree of the Queen. It is to a world like this that the scientist must conform in totalitarian countries, no matter whether they be those of the right or of the left. The Marxist Queen is very arbitrary indeed, and the fascist Queen is a good match for her.

What I say about the need for faith in science is equally true for a purely causative world and for one in which probability rules. No amount of purely objective and disconnected observation can show that probability is a valid notion. To put the same statement in other language, the laws of induction in logic cannot be established inductively. Inductive logic, the logic of Bacon, is rather something on which we can act than something which we can prove, and to act on it is a supreme assertion of faith. It is in this connection that I must say that Einstein's dictum concerning the directness of God is itself a statement of faith. Science is a way of life which can only flourish when men are free to have faith. A faith which we follow upon orders imposed from outside is no faith, and a community which puts its dependence upon such a pseudo-faith is ultimately bound to ruin itself because of the paralysis which the lack of a healthily growing science imposes upon it.

INDEX

This edition of
The Human Use of Human Beings: Cybernetics and Society
was finished in March 1989.

It was printed on a Crabtree NP 56
on 80g/m^2 vol. 18 book wove.

The new material for this edition
was commissioned and
edited by Les Levidow,
copy-edited by Selina O'Grady
and produced by Martin Klopstock and Selina O'Grady
for Free Association Books.